GROWING
PLUMERIAS
IN HAWAI‘I
AND AROUND THE WORLD

JIM LITTLE

Mutual Publishing

Library of Congress Catalog-in-Publication Data

Little, Jim, 1937-
    Growing plumerias in Hawai'i and around the world / Jim Little.
        p. cm.
Summary: "Over thirty years of research, discovery, and experimentation are presented in this reference guide essential for plumeria enthusiasts interested in growing its trees. Information about different types of plumeria, asexual propagation, rooting and media structures, cultivation, pest and weed control, hybridization, cross-pollination, collecting, and tips on plumeria care are discussed. Special diagrams and full-color images included"--Provided by publisher.
    Includes bibliographical references and index.
    ISBN-13: 978-1-56647-772-7 (pbk. : alk. paper)
    ISBN-10: 1-56647-772-7 (pbk. : alk. paper)
    1. Plumeria--Hawaii. I. Title.
    SB413.P56L58 2006
    635.9'3393--dc22
                    2006005485

ISBN-13: 978-1-56647-772-7
ISBN-10: 1-56647-772-7

All photos are taken directly from the JL farms without any digital enhancement or manipulation
Photography by Jim and Clark Little
Image on cover: Waimea, photo by Clark Little
Image on title page, iii and iv: JL seedling
Portraits on page vii submitted by Dr. Richard Criley,
Bud Guillot, Paul Weissich, Jim Little and Betty Gibb

Illustrations by Jared K. Young
Design by Emily R. Lee
Contributing editors: Dr. Richard Criley, Dr. Doric Little, and Carolyn Rice

Second Edition, Third Printing, January 2008
            Fourth Printing, April 2010
            Fifth Printing, October 2012
            Sixth Printing, October 2015
Mutual Publishing, LLC
1215 Center Street, Suite 210
Honolulu, Hawai'i 96816
Ph: 808-732-1709 / Fax: 808-734-4094
Email: info@mutualpublishing.com
www.mutualpublishing.com

Printed in Korea

TABLE OF CONTENTS

ACKNOWLEDGMENTS

It is with a great deal of excitement that I share information on growing plumerias learned from over thirty years of experience and discovery. Customers here and around the world have shared with me their growing experiences, and their lessons learned will be discussed. People who have been instrumental in my development and to whom I am most indebted are the support of my loving family: my wife, Doric, who has supported me in all my endeavors during our forty-six years of marriage; my son Clark who has been at my side from the beginning and whose extra senses enable him to observe subtle differences in flowers and to visualize their photographic possibilities; and to my son, Brock, a professional big wave surfer and now stunt man who enjoys volunteering my flowers for any special occasion.

I am further grateful to the many wonderful people who have guided me to my current position in the plumeria world. My first mentor was Donald Angus (1908-2001), a philanthropist who traveled the world collecting plumeria cuttings and seeds. Soon after Donald developed my interest in this very special flower, Dr. Richard Criley, one of the foremost authorities in the world on plumeria and tropical plants, began sharing his vast understanding with me. Adding to my appetite to learn more were Bill Moragne, father of modern plumeria (1905-1983) of Kaua'i who taught me the techniques of hand pollination and shared his crosses with me, Ted Chinn, Donald Angus, Dr's. Richard Hamilton, Donald Watson, James Brewbaker, and Horace Clay who, among others, were

Dane JL seedling

instrumental in collecting the original sixty-four plumerias that started the University of Hawai'i experimental farm in 1964. Ted Chinn, who served under the tutelage of Dr. Richard Criley in 1968-1969, used this germplasm for the study and research of plumerias. About the same time Paul Weissich, former director of Honolulu Botanical Gardens, had the vision to establish a plumeria garden in Koko Head Crater on the east side of the island of O'ahu.

Still others include Richard and Mary Helen Eggenberger and Elizabeth Thornton (1919-2005). I am indebted to the Eggenbergers for developing a business (The Original Plumeria People) to export plumerias from Hawai'i to Texas, which, in turn, propelled an interest in plumeria collecting worldwide. I am also grateful to Elizabeth Thornton who with other Texan pioneers, Nancy Ames and Nadine Barr, started the Plumeria Society of America in 1979. In 1983, I had the pleasure of meeting Elizabeth in Hawai'i where we were joined by Donald Angus to discuss the business of exporting plumerias to Texas.

Other acquaintances with whom I have exchanged information include: Bud Guillot, who has exquisite taste in collecting one-of-a-kind plumerias and not duplicating or renaming existing cultivars; Emerson Willis, who for years has been a true ambassador in his travels to collect, trade, plant, and promote plumerias; Milton Pierson, who relentlessly has collected, studied, and fielded hundreds of questions from individuals; and Henry "Apples" Dupree (1909-2002), with his untiring efforts to plant plumeria trees in parks, city, government, and military installations. Henry was instrumental, together with Carl and Joy Herzog, George Emerick, Venessa Marlow, and Margaret Dunivan in starting the Southern California Plumeria Society, which incorporated in 2000. His dream was to see a plumeria tree planted in every yard in San Diego. Other devoted people are active in plumeria societies in the United States; including: the South Coast Plumeria Society of Huntington Beach, California, the Valley of the Sun Plumeria Society of Arizona, the Gulf Coast Plumeria Society in Florida, and the Plumeria Society of South Texas in Corpus Christi.

Further, I am grateful to Kampon Tansacha and Michael D. Ferrero of Nong Nooch Gardens in Thailand who openly traded rare and limited plumerias with me, and David Orr of Waimea Audubon Botanical Garden, who has been so kind in sharing plumeria species and botanical information with me. I am thankful to countless others who have enriched my knowledge by sharing their findings and cuttings with me.

Finally, I will forever be indebted to the Waialua United Church of Christ for subleasing me two acres of land to start my first plumeria farm. This gave me the opportunity to use new land that led to the largest research and acquisitions plumeria farms in Hawai'i and, perhaps, the world.

Row 1

Row 2

Row 3

Row 4

From top:
Row 1: Bill Moragne; Bud Guillot; Donald Angus; Row 2: Henry "Apples" Dupree; Allison and Clark Little; Elizabeth Thornton; Milton Pierson; Row 3: Dr. Richard Criley; Jim Little, Ted Chinn, Richard Criley; Paul Weissich; Row 4: Ted Chinn with Donald Angus flower; Emerson Willis; Mary Helen and Richard Eggenberger; Bottom Right: "Future plumeria leaders" Allison, Sandy, Dane and Clark

FOREWORD

My inspiration for writing this book started in 1974 when my customers began asking me questions on how to best cultivate plumerias. The questions have come from collectors here and abroad and from the new generation of plumeria fanciers who have formed a bond with this remarkable tree. It is my hope that your questions will be answered and you will quickly develop an appetite to read on and learn more about this easy-to-grow exotic plant.

Feeding this growing frenzy are the energetic plumeria societies introduced in Texas, California, Florida, Arizona, and Australia. Commercial backyard nurseries and private venders, hobbyists, email groups, and chat sessions are all adding to the knowledge base about plumerias. People in the mainland United States, Hawai'i, and countries from all over the world, notably Thailand, Singapore, India, Australia, Indonesia, and Malaysia are sharing their enthusiasm for growing plumerias.

This book is written for all levels of plumeria growers, including the experienced collector who wants to sharpen his/her skills as a grower. The book will highlight growing techniques and share with you the best methods for successful plumeria cultivation as well the latest developments in new cultivars, trends, tricks, and surprises. In this second edition you will be introduced to one of the fastest growing gardening trends in America today, the plumeria, and to the possibilities for growing them around the world.

This overview will illustrate with text and photographs (many never seen before) the recent developments in growing plumerias.

JL Red-orange

Plumeria blooming in planters on rooftops in Guangzhou (formerly Canton) China

JL Miniature

JL Miniature

GROWING PLUMERIA AROUND THE WORLD

NEW EXPANDED SECOND EDITION: THIRD PRINTING.

The formula for growing plumeria "Anywhere in the World" is the same for growing plumeria in Hawaii, California, Texas, Florida, or in any tropical or subtropical state or country. As long as there is a half-day of sunlight, plumeria will grow, providing the prevailing temperatures are above forty degrees. When the "Queen of Plumeria", Elizabeth Thornton wrote her first book "The Exotic Plumeria" in 1978, she never dreamed that her introduction to growing plumeria in pots would lead to one of the most popular plant hobbies in the world.

Every plumeria cultivar is different and some tolerate the cold and light better than others. Many collectors of plumeria who live in colder geographical locations with lower sunlight claim that they can grow these exotic plants. They have learned that variables within their microclimate can influence their success or failure. Individual microclimates are influenced by temperature duration, sun direction, shade, prevailing winds, altitude, moisture, surrounding shrubs, trees, and buildings, any or all of which can help insulate against coldness. The main difference is that collectors must adjust their growing techniques to a shorter season and experiment to find what plumerias are the best varieties for the lower temperatures and shorter days of sunlight in their microclimate.

More people are experimenting and growing plumeria in all parts of the world. They include commercial and individual collectors in many parts of Europe, the Middle East, South Africa, Japan, Australia, India, the United Kingdom, Canada and now, China. They are experimenting with different cultivars, heating conditions, artificial lighting, greenhouses, sunrooms and sunlight penetrating through windows in their homes. Those growers who have the patience to develop their learning curve can grow plumeria or most any tropical or sub-tropical plant. There are many plumeria grown under less than ideal conditions, which suggest a patient grower educated with the proper knowledge, can successfully grow plumeria.

The process begins when a rooted cutting is acquired from a trustworthy source. Depending on the cultivar and time of year the cutting was taken and rooted, the plumeria should blossom within the first or second year. While unrooted cuttings can be rooted in colder areas, it takes longer unless you can provide: bottom heat (heat mat), a heated greenhouse, an overhead misting system with warm water or overhead lighting designed for growing plants. You can use one of the above techniques or all of them combined.

Some plumeria begin to falter at forty degree temperatures while others will take a brief freezing with no setback. There are no definitive studies to determine which varieties resist the cold best. Plants in pots offer better protection from the cold than those planted in the ground. Growers can put two pots together to insulate for extra protection.

Fresh plumeria seeds can be germinated at anytime and grown throughout the year in a warm environment. The growing period will be slower during the winter months, but watching the seedlings develop into mature plants makes the experience satisfying. Anticipating the surprise of a new flower adds to the pleasure.

When the temperatures begin to dip into the low forties, take some old flannel or wool clothing and wrap the pot and trunk of the tree. Bring the plant inside and keep it in a warm place until springtime temperatures warm up. While it is not required, a plumeria is happy to be misted occasionally during its dormancy without watering the medium. When temperatures begin to elevate in the spring, remove the flannel; water lightly and liquid fertilize gradually increasing the dosage as new foliage appears. More information regarding this topic can be found under "Winter Protection" section on page 45-46.

It is not known why some cultivars offer more protection against colder temperatures. Celadine is one known cultivar that some customers have experienced as being more cold tolerant. To recommend specific plumerias for an area where so many isolated cold climate variations exist is difficult. Someone in a cold pocket in Seattle or Germany could have excellent winter tolerance while another person in the same cold temperature zone in Arizona or Canada could fail. This type of inconsistency has proven true for growers comparing growth habits in tropical or subtropical microclimates. Much more is to be learned about climate variations.

JL seedling

JL Seedling

Cindy Schmidt

JL Miniature

JL Seedling

P. *obtusa* ssp. *sericifolia* var. *tuberculata*

Paul Weissich

Hawai'i is home to some of the most exotic plumerias (also known as frangipani) found anywhere in the world. Plumerias belong to the Apocynaceae family, which is composed of approximately 170 genera and 2,000 species. There are about twenty genera to be found in Hawai'i as cultivated ornamentals.

The blossoms of the plumeria tree are certain to awaken your senses. You will be immediately attracted by their colors, fragrances, shapes, and foliage. During the winter months the leaves will drop to the ground and the flowers cease to bloom as the trees rest. Predictably, every spring, the skeletal forms of these deciduous beauties return from hibernation and explode with an array of colorful and fragrant blooms.

The genus *Plumeria* was named to commemorate Charles Plumier, 1646–1706, a French botanist who traveled the world on botanical excursions searching, collecting, and studying plumerias along with other floral interests. When the first yellow-white flowered tree was introduced to Hawai'i by Dr. William Hillebrand in 1860, it became a celebrated tree, popular in Hawai'i's landscaping. A cutting from this original tree can still be seen at Foster Botanical Garden in Honolulu.

It became known as Graveyard Yellow or Temple flower as it was planted in cemeteries as a symbol of eternal life. It is also known as Common Yellow or by the name Celadine. The term Graveyard Yellow has lost its reputation in

Celadine plumeria tree—
Honolulu Botanical Gardens

Celadine accession label—
Honolulu Botanical Gardens

Celadine inflorescence

modern times. This stigma has been forgotten by today's collectors who show no interest in old plumeria stereotyping.

In 1931, Harold Lyon, for whom Lyon Arboretum has been named, introduced the first Singapore plumeria to Hawai'i from the Singapore Botanical Gardens. It is correctly referred to as *Plumeria obtusa*, an evergreen tree that retains most of its leaves throughout the year.

Singapore obtusa against Mt. Ka'ala

Singapore obtusa trees at the University of Hawai'i Mānoa Campus

At about the same time, a red flower was introduced into Hawai'i from Mexico. Speculation gives credit to either Mrs. Paul Newmann, wife of a consul general in Hawai'i, or Mr. Gifford, a landscaper for the Royal Hawaiian Hotel. Donald Angus gives credit to Kauka Wilder, an associate botanist at Bishop Museum who traveled extensively through Mexico.

The forms *Plumeria acutifolia* and *acumulata* were previously thought to be separate species. According to George Staples, in *A Tropical Garden Flora*, (page 124–125), we can correctly refer to the yellows, whites, reds, and rainbow colors as species *Plumeria rubra*.

Today, the combination of natural hybridization with some deliberate cross pollinating has led to countless numbers of hybrid cultivars. These exotic cultivars have promoted a resurgence in plumeria popularity that is unstoppable. Farmers in Hawai'i who reduced their tree count in the 1980s and 90s are now replanting to meet the requests for plumeria flowers, landscaping, and the new and growing demand for exporting plants, cuttings, and flower leis.

"Most young people find botany a dull study. So it is, as taught from the text-books in the schools; but study it yourself in the fields and woods and you will find it a source of perennial delight."
CHARLES PLUMIER

"For the amount of care expended the plumeria it will give a ten-fold amount of pleasure. The flowers stay fresh and fragrant for days when brought in the house and placed in water, and continuous scented blooms will glorify your garden for months."
ELIZABETH THORNTON

"It is only a matter of time before plant collectors everywhere will have a plumeria growing in their landscape or as a potted plant in their house."
JIM LITTLE

Form

The Plumeria is considered by many to be the most celebrated of all flowering trees growing in the tropical and subtropical world today, and is one of the most engaging, appealing, and captivating trees in existence. Its broad appeal is best characterized by its inflorescence (flower cluster), flower types, size variation, colors, fragrances, and growth habits, and, with all these variables, any combination can occur. The peak flowering period for most plumeria trees around the world is from April through September. Few other trees give you as many different looks as a plumeria. In addition, the tree takes on many sculptural shapes depending on the cultivar or species.

From top to bottom: Attractive wavy petals; red and pink banding; greenish florets developing into yellow flowers.

Top: Compact inflorescence
Bottom: Floral envelope with red band

Contrasting flower colors

The branching habits and the different leaf shapes and shades of green add to the beauty of this deciduous plant. Leaves vary in shapes, tips, textures, sizes, color shades, margins, and veining. Trees are used to beautify the setting of a home, landscape large developments, or to serve as a showplace for a potted plant on a lānai of an apartment or condominium.

Most plumerias grow well in containers and can be rotated inside a home for short periods of time. Plumeria collectors can take a small space and create a subtropical garden around their home using the different growth habits of the tree. Tree habits can spread horizontally, hanging from terraces, or grow vertically, like a cactus, emphasizing dramatic shapes. Plumeria flowers can show off their brilliant colors against a backdrop of foliage to highlight a garden setting or can be picked to make a lei, worn in the hair, made into wrist bands, floated in center pieces, or used for ceremonial decorations. The creative ways to use the flowers are endless.

Every year this deciduous tree will lose its leaves and go to sleep for several months as winter approaches and temperatures dip. It will reawaken in the early spring as the branch terminals start to turn a bright waxy color and initiate new leaf growth. This is the first sign telling you to prepare to repot, relocate, and anticipate another season of flower charm and fascination (see Golden Pagoda pictures on page 11 for color comparisons, variations, and off-season trees).

Color

The colors of the plumeria flower are among the most popular attractions for plumeria collectors and, with the development of new hybrids, the choices are endless. The color evolution developed from the common whites, yellows,

Kimo

pinks, and reds into a rainbow of trendy variations. Color hues are intense, dull, spotted, streaked, splashed, muted, and multicolored. The flowers can exhibit changes in color depending on the time of day, light intensity, adaptation, time of year, location, age of the flower or tree, culture, micro-climate, humidity, and cloud cover. The inflorescence may vary from a few flowers to a large cluster of flowers with a short or long flowering duration. Yearly flower abundance and quality can change depending on nature's influence.

JL Dwarf Pink pudica Seedling

Plastic Pink Same flower; different color created by sun bleach and aging flower

Lava Flow

Bali Hai with no red banding

New color breakthroughs come in lavender, purple, blue (a little), green (a little), silver, salmon, rust, coral, burgundy, chartreuse, fuchsia, and many combinations with different color veining streaking through the blossoms (see pictures of flowers that show these dramatic changes). While claims have been made, there is no full-bodied black, blue, or green flower at this time. Full colored orange hues, which are rare, are very close to becoming a reality.

Shani orange pastel with cupped elliptic petals

Black Ruby rare black veins in red flower with black eye

Cherry Blossom Pink JL seedling, soft pink with yellow eye

9

Flowers begin to appear in the spring and continue into November or later depending on the tree and geographical area. Color shades and flower shapes can change during this period especially when flowers are compared from state to state and season to season. Photographic variables that determine exposure can lead to further confusion when comparing a flower in Texas with the same variety in California, Florida, or Hawai'i. Some varieties are early, mid or late set bloomers, while others will bloom for the whole season.

There can be many seasonal looks to a plumeria that can add confusion to a named cultivar. As the season comes to an end, usually around October, some flowers are smaller, disfigured, not as intense in color, and have a diminished flower quality. It is known that a few trees will flower into early December. *Plumeria pudica* (Bridal Bouquet), for example, is a long-season bloomer. Off-season trees like San Germain and Hong Kong will come into bloom while others are ready for winter hibernation.

This rare orange flamed plumeria was named for the author's wife, Doric.

Seasonal change comparison in Golden Pagoda, JL Seedling

My Valentine red band on front and back of flower

Flower Petals

Plumeria flowers, which averaged two-and-half inches in diameter, are now compared with new cultivars ranging from five to seven inches and bigger. This isn't to say that a two-inch flower can't be just as captivating as a seven-inch flower. In fact, some collectors seek out tiny flowers. Flower petal shapes can be wide, narrow, overlapping, twisted, reflexed, wavy, rounded, or with pointed tips.

Daisy Wilcox with Keiki Lavender—
size comparison

Keiki Lavender with red leaf veins

Folding petals with white underside

FLOWER TYPES

shell-type
(usually does not open further)

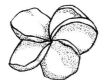

wide petals
rounded tip
moderately overlapped

obovate petals
rounded tip
slightly overlapped

wide, obovate petals
rounded tip
moderately overlapped

wide petals
pointed tip
moderately overlapped

wide, elliptical petals
rounded tip
highly overlapped

wide, elliptical petals
pointed tip
moderately overlapped

elliptical petals
pointed tip
moderately overlapped

elliptical petals
pointed tip
slightly overlapped

narrow petals
pointed tip
slightly overlapped

narrow petals
pointed tip
slightly overlapped

narrow, twisted petals
pointed tip
slightly overlapped

narrow petals
tapering tip
no overlap

Assorted flower colors and shapes

Front and back of the same flower   Backside variations

Backside variations

Often overlooked is the underside of the flower petal, which, when observed, can be just as striking as the face of the flower. This underside is often prominent when the flowers are made into a lei.

Many new features are being developed from thousands of seedlings grown by collectors worldwide. While there is no guarantee that an open pollinated seedling (natural cross vs. deliberate cross) will produce a superior cultivar, exceptions do occur. Flower petal variations are evolving not only in size and color, but in fragrance and texture as well. Petals vary from paper thin to leather thick, each with a different sheen. Some have a high gloss surface while others have a matte finish. Another eye-catching feature is the peduncle, the stalk that supports the inflorescence, which takes on different colors, textures, and configurations highlighting the beauty of the inflorescence. Plumeria flowers can

From top left clockwise: two examples of compact inflorescence; all red petals translucent against the blue sky; Lutica; and Giant rose shell

be defined in many ways and one of the most important features is the fullness and lasting quality of the inflorescence. Those with a large inflorescence can be enjoyed for days as the buds continue to open.

Flower longevity has been tested for use in lei making. Flowers like Celadine, Pops Red, Nebel's Rainbow (a.k.a. Lei Rainbow), Donald Angus, Penang Peach, and Candy Stripe have proven to have excellent keeping qualities, sometimes up to three days, depending on the care of the lei maker. A new favorite, Vera Cruz Rose, does not last as long but its popularity is due to its aromatic rose fragrance. To increase the longevity of the flowers, lei pickers harvest Vera Cruz Rose when the flowers are newly opened, still in a shell form, to create a beautiful and lasting lei.

Vera Cruz Rose has a sweet rose fragrance

Fragrance

The scent of a Plumeria is considered to be so diverse that almost everyone will discover a fragrance that pleases his/her olfactory senses. Each person has his/her own definition of fragrance appeal and sensitivity to smell. Fragrances abound as frequently as different colors. Most flowers are pleasantly fragrant but some can smell like a dirty dog.

Vera Cruz Rose inflorescence

Pleasant aromas that resemble everyday associations include soap, flowers, roses, gardenias, rain, detergent, fruit, ginger, grapes, gum, lotions, coconut, candy, citrus, fresh air, spices, and hundreds of other different variations. No other flower gives you so many fragrant choices.

Vera Cruz Rose, which has a fragrance as sweet as a rose, and San Germain, two highly scented trees, are considered to be among the top choices.

Plumeria fragrances are capped in a bottle and sold as perfumes and used for aromatic pleasures in soaps, candles, and potpourri. Shades of white, pink, and yellow flowers are generally more aromatic than the commonly thick-bodied reds and some rainbows. The true flower fragrance is usually not derived from plumerias but from chemicals. According to Dr. Criley, "There can be 60+ volatile compounds that make up a natural plumeria fragrance and most of these are not included in the chemist's formulation of a plumeria perfume."

Tree habit and Leaf shapes

When stems are cut, leaves broken, or flowers harvested, a milky latex which contains some resinous rubber is exuded. It is known to have some medicinal properties when it is used as a purgative, but can be toxic in large doses. The foliage takes on many shades of green, sheen, and shapes which vary from curly, pointed, large, small, veining, glossy, dull, and petiole (stalk of the leaf) features.

Some collectors are first attracted by their love for foliage. Variations abound in this category and when the flowers are not always present the foliage continues to stand out (see illustrations on leaf shapes and include leaf tips, bases, and margin veinations).

Oblanceolate leaf with obtuse leaf tip

Green peduncle supporting inflorescence

Leaf with wavy margins

Natural drooping leaf habit

Leaf Shapes

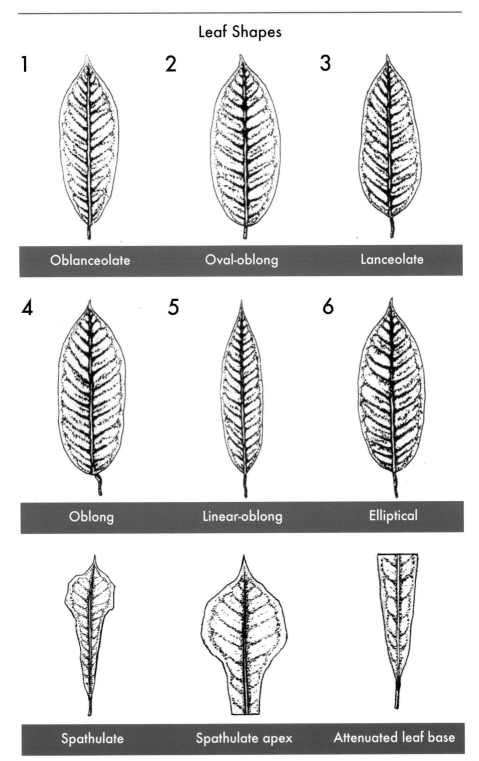

1 Oblanceolate

2 Oval-oblong

3 Lanceolate

4 Oblong

5 Linear-oblong

6 Elliptical

Spathulate

Spathulate apex

Attenuated leaf base

Acute Leaf Tips

| Broadly | Medium | Narrow | Obtuse | Rounded |

Acuminate Leaf Tips

Leaf Margins

| Abrupt | Broadly | Narrow | Spathulate | Spathulate apex | Attenuated leaf base |

Variegated Samples

Ten year old JL Dwarf Singapore Obtusa

Top left: Deciduous elliptic leaf shape with acuminata-acute leaf tip; lower left: evergreen elliptical leaf shape with obtuse leaf tip and green; right: glossy Singapore leaves

Plumerias have many growth habits and can grow horizontal to the ground or upright. Tree forms are symmetrical or asymmetrical and can grow lanky or compact, twisted or elongated depending on the cultivar. Some have branches the size of oak trees while others have pencil thin branches by nature. Branch textures can be corky, have needle-point roughness, or be glide smooth. Branching features include multiple branched, fascinated, weeping, prostrate, upright, and tuberculated. You will be surprised at how many features there are to plumeria trees.

Clockwise from bottom left: P. *obtusa* ssp. *sericifolia* var. *tuberculata*; elongated growth habit in early summer; elongated branching habit in early fall; orange-red peduncle with dark red pedicels; tubercules variations

Clockwise from bottom left: Lurline Matsumoto; multiple branching from single trunk; Kāneʻohe Sunburst with naturally thin branches.

23

Seeds

Using seeds continues to be a very popular method of propagation. Plumeria seed follicles come in many different widths, lengths, shades of brown and green, and can hold twenty-five to one hundred seeds per pod, dividing in to two pods (infrequently one) as the seeds mature. It is not unusual for some trees to self-pollinate and produce up to fifty or more follicles in a season depending on the cultivar and age. When the seedpod has fully matured, usually around nine months, you will first notice some shriveling as it turns brownish-black and starts to split open exposing the seeds. The seeds are now most viable and can be removed and planted.

The first step is to gently feel the seeds for a cotyledon and discard any that feel flat or have no evidence of an embryo. Next, lay the seeds on a loose media,

Clockwise from bottom left: Unripe green seedpod; forming stages of young seedpods; four- to nine-month growth period; newly opened seedpod.

perlite or other light-bodied rooting mix, and cover with a thin layer of perlite. The seeds do not have to be standing upright in the mix as they will turn their bodies to the light no matter how you place them. Fresh seeds should germinate in twelve to fifteen days. While many types of seeds can be presoaked for a few hours or overnight, this procedure is not recommended for plumerias. Plumeria seeds are soft bodied so damping-off from overwatering, caused by certain fungi that prefer moist environments, is often the main cause of failure. Soaking seeds in water overnight is recommended for hard-shell seeds, such as palm seeds, but not plumeria seeds.

Perfectly formed seeds ready to plant

After the seeds have germinated, you can start feeding them with any balanced liquid fertilizer at half strength every ten days. Within the first year, the seedling will have developed a healthy vegetative growth and it will be time to change to an 18–6–12 NPK ratio. Contrary to popular belief, a higher nitrogen fertilizer is preferable to a phosphorus fertilizer in inducing first flowering. New studies (see Flower Tech in references) show that young plants do not need so much phosphate. Phosphate is recommended for mature plants that have established root systems.

When seedlings reach two to three inches, they may be transplanted into a four-to-six inch pot using a well-drained potting mix and then progressively replanted as they outgrow their pot size. Generally speaking, a 90 percent germination rate or better should result if the seeds are fresh. Seeds will stay viable for up to a year and sometimes longer if they are held indoors in low light with an average temperature. Paper bags should be used for seed storage, not plastic baggies, as the seeds need to breathe. Seeds may be refrigerated, but this is not necessary. The older the seeds, the slower the germination, and the higher the attrition.

It takes an average of eighteen to thirty-six months for the plumeria seedling to develop its first flower. First flowering will occur sooner if the seedling is grown in the ground rather than a pot. It has been recorded that some have waited for fifteen to twenty years for their first flower. This, however, is rare. The worst-case scenario is that it could be an inferior first flower not worth keeping at any age.

The first flower is not always indicative of the true and consistent look that will follow. In some seedlings, it takes three to five years to see if the flower color and shape remain constant, particularly in more exotic cultivars. Pinks, yellows, and whites generally remain constant from the first flowering. Rainbow hues offer the greatest color irregularities. Flower shapes and sizes can also vary year to year.

The process of seedling selection is very important if you are interested in perpetuating only the best flowers. Waiting is often very difficult. Picasso once said about his paintings that he kept only one out of every hundred. The plumeria fancier should be as discriminating.

Many seedlings will be similar in shape and color. Seeds of white and pink parents are most likely to give white or pink progeny. Seeds from rainbow trees produce the most original seedlings. This doesn't mean, however, that a seedling from a white parent can't throw a surprise rainbow color. Relying on an open pollinated seedling to be original is like putting your money on the roulette wheel. The chances of hitting the jackpot is limited only by the number of plays you make. However, the chances of good open-pollinated seedlings are enhanced if surrounding plumeria trees are of mixed colors.

The best chances for exciting seedlings are to practice proven hybridizing methods and, above all, be patient! Hand pollination is very time consuming. To

increase the chances of potentially good seedlings without the required dedication, plant open-pollinated third and fourth-generation seeds that come from exceptional flowers. Good flowers come from exceptional seedling parents and can produce a very valuable offspring. Contrary to popular belief not all cultivars produce seeds (see hybridizing).

Above left: Newly planted plumeria seeds; below: young tree with many seed pods

Cuttings of different lengths and diameter

Cuttings and Rooting

Based on the author's experience, good cutting requires a sharp pair of hand shears to make a clean diagonal or perpendicular cut. It is recommended to dip or spray the shears with rubbing alcohol or peroxide to kill any bacteria prior to making the cut. When making a cutting, select a branch cutting from the tree that looks robust and with as little new green tip growth as possible. If you prefer low branching, select a cutting with a short axis and multiple tips. Flower farmers use this technique for field trees and ease of harvesting flowers. Cuttings ordered by mail or cut in the early spring will stay hardy and be ready to root after they have hardened off, even if you delay the rooting process for several weeks. Questions on whether to cut off an existing flower spike (peduncle) before rooting are often asked. Only new barely visible emerging peduncles should be left to develop, but the rooting time will be slightly longer. The first flowers, however, will be smaller.

To reduce transpiration, at anytime of the year, remove all the leaves by cutting the petiole (leaf stem) close to the branch except for the terminal leaf. In a few days, the cut petioles will fall off and this is a sign that you may begin the rooting process. If you strip the leaves by hand, it can increase the chances of leaf node disease and decay. After the cutting is taken from the tree it is best to wait several days (seven to fifteen days or longer depending on the cultivar), for the soft base of the cut surface to form a callus (hard to the touch) before beginning the rooting process. Hot waxing the cuts will help reduce disease and speed rooting for hard-to -root cuttings, but is not necessary unless you want to minimize all variables that can lead to cutting failures.

Regardless of where plumerias grow, cuttings can be taken in sizes from as little as three inches to seven feet long, in any diameter, and at any time of the year. There are few trees that are as resilient to constant pruning of its branches as the plumeria tree.

Above: Plumeria grove looking toward Mt. Ka'ala with a row of Grace dwarf trees
Below: Ten-year-old Dwarf Petite Singapore

Rooting time will vary depending on the varieties, branch maturity, and rooting technique. Branch density varies from very soft to hard wood. Some cuttings will stay viable for the rooting process up to three months or longer depending on the time of year the cutting is taken and the selected cultivar.

Although cuttings can be taken any month of the year, plumeria will root faster and flower sooner if taken in the early spring or when the tree is still in dormancy. It is best to avoid propagating cuttings when the tree is in its repid growth state.

To root, simply plant the cutting four-to-eight inches deep depending on the size of the cutting into a rooting mix, (see pages 36-37 on rooting) using any container or poly bag with good drainage, and place it on the warm concrete or ground (tests have proven that a translucent container induces faster rooting). It makes no difference if the cutting leans diagonally against the side of the pot while rooting. The heat generated from the ground surface will hasten the rooting time. Using the warm water from your hose will also speed the rooting process. Heat mats, misting with a fine water spray (by hand or automatic timers) and hormones also help speed rooting times. More people are beginning to use halide heat lamps (400 watts is recommended) to initiate rooting and force flowering.

New studies are now focusing on geothermal heat for rooting applications. Regardless of the choices for rooting cuttings, learning by practice what works best will ultimately lead to success.

A cutting can be speed rooted using the JL method described below.

Rooting a cutting directly in the ground

29

JL prehardening method. Left: Cut the leaves off, Middle: Notch out a section. Right: Cut branch diagonally without severing it completely.

Simply precut or notch out a section of the branch without severing it completely from the tree and let it hang naturally clinging by its epidermis (bark) for a couple of weeks or indefinitely (see photo). After the cutting has hardened and callused on the tree, you can cut the branch anytime to start the rooting process immediately. A rooting hormone or basal dusts with fungicides would aid in the development where rooting conditions are not ideal.

Gang rooting is a very popular method for rooting several cuttings at the same time. Plant approximately ten hardened cuttings into one large pot following the rooting instructions. When rooted, usually in about three months, a slight tug on one of the cuttings or visible roots at the drainage holes will indicate the roots have developed and new bright shiny leaf tips begin emerging. This is a sign for the cuttings to be divided and transplanted into single pots. Tap the bottom of the pot, remove the cluster of rooted cuttings, divide, and repot them into six-to-ten inch pots depending on the size of the newly-rooted cutting.

If the cutting is fully callused, rooting in water with a mild liquid fertilizer can be done, but it is not recommended as the cutting must adjust from a water root structure to a soil mix structure. Furthermore, different internal root structures that develop in water do not provide the fine hair roots to adapt to a new rooting mix. While the cutting may look all right

Gang rooting dwarf tips

for a brief time, the main roots without the hair roots will sometimes weaken and die from lack of oxygen. An exception to this finding is that sometimes older dehydrated cuttings can be soaked in water for several days before starting the rooting process.

Bag Rooting

A less conventional experimental method is rooting cuttings in sealed plastic bags. This method requires consideration of the following factors: age of cutting, time of year, plumeria variety, cultural medium, moisture content, environmental conditions (light source, bottom heat and temperature), rooting hormones and control of pathogens. Both the Air Layering and bag rooting technique use the plastic enclosure for rooting. The significant difference is that the air layering method is performed on a living branch of the tree which most often leads to successful rooting.

Cuttings may also be started from as small as three to six-inch tips or mid-stem branches. They must be laid flat in a pot or tray. This rooting technique is slow and sometime the attrition is high, but is done by nurserymen who know the tricks and practiced by people who like to multiply cuttings into thousands of logs for future uses.

The earlier the cuttings are taken starting in January (even if you protect them until springtime before rooting), the faster the rooting process. Cuttings are slower to root if they are taken towards the end of the flowering season, normally after August or September. If cuttings are taken at the end of the growing season (November or December), it is best to store them inside or plant them in perlite and hold them indoors until the following spring. Available lighting or heat lamps and very light watering, with dry periods in between, or misting are recommended if you choose this option. The main factors that determine the time it takes to produce a first flowering are the age of the cultivar, maturity of the branch cutting, and month of the year the cutting is taken.

Rooting in a black bag. Photo by Dr. Richard Criley

Three Bali Whirl ten-petal seedlings

Grafting

Grafting plumerias is not necessary and only recommended for very hard to root cuttings and for saving the tip of a dying plumeria. Even then, the consequence of the graft is unpredictable. There is no proven plumeria seedling rootstock upon which to graft scion wood. Every plumeria seedling can be different, even from the same tree and same seedpod. The scion wood can be the same across a number of grafts, including dwarfs, on the same seedlings. The difference is that root systems could potentially offset scion growth. In addition, taproots which develop from seedlings are not recommended for container stock, as the roots will coil and retard the growth which could affect the flowering of the plant.

There have been years of research on selected ornamental, nut, and fruit trees to determine the most consistent root stocks for grafting. Because plumeria grafting is relatively new, we have no way of predicting consistent results of a potential graft until studies can qualify rootstocks compatible for all types of scion wood. Because this particular research has never been done with plumerias, the scion wood and grafting stock will continue to be incompatible with many grafts. Current grafting methods will often lead to cambium mismatches which are apparent in different plumeria cultivars and species, due mainly to inherent physical differences. It is not known what will be a homogeneous mixture of the scion and grafting stock. This is especially true for retaining the natural beauty of a dwarf or species plumeria.

Exceptions can be found as a few people have created some success by grafting onto rooted cuttings from selected cultivars rather than seedlings. This technique would also eliminate the taproot that develops from seedlings. This would require study to determine which trees are best for this procedure. Commercial nurseries, limited by land space, use grafting to multiply their stock of plants. If large trees are available for cuttings, it is not necessary to graft. For aesthetic purposes and to remove any doubt, plumeria fanciers prefer a nongrafted plant.

If one likes to experiment it can be challenging to graft multiple colors onto a rooted cutting or tree. This necessitates deciding on the scion color mix and the wood stock. Some trees will accept different color grafts, but often one of the scion grafts or the stock (mother) plant will dominate and spoil the multicolor balance. There are new experimental grafting methods and techniques along with chemical mixtures and glue guns that can be used in treating the surface of a graft that will induce a faster callus formation and fusion of scion and root stock.

Grafting is a technique that takes much practice to achieve high rates of success. Cutting propagation is a simple practice with fewer failures. While a grafting knife for right- or left-handed people is most often used, special grafting

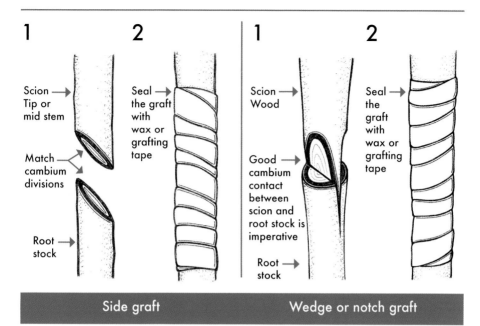

1

Scion →
Tip or
mid stem

Match →
cambium
divisions

Root →
stock

2

Seal →
the graft
with
wax or
grafting
tape

Side graft

1

Scion →
Wood

Good →
cambium
contact
between
scion and
root stock is
imperative

Root →
stock

2

Seal →
the
graft
with
wax or
grafting
tape

Wedge or notch graft

sp. bahamensis considered by many to be one of the rarest plumerias

tools can be purchased for creating the perfect cut. The grafter should consider what types of grafts work best for plumerias and what grafting methods to use. Each grafter develops his/her own favorite method. The side and the V-shaped wedge graft are the most popular methods. The key to successful grafting is to match the cambium layers of the scion wood and stock. Regardless, the same grafting consequences mentioned above still exist.

All things being equal, over a period of several years, a rooted cutting will outperform a grafted tree in terms of flower size and regularity of quality seasonable blooms. If grafting is your choice, it will require maintenance, until convincing data can assure that these uncertainties can be corrected.

The biggest and most beautiful trees seen growing in Hawai'i, California, Texas, Florida, and subtropical parts of the world have not been grafted. Learning the proper techniques of rooting a cutting will avoid your ever having to graft. Grafted plumerias will often look like the pictures seen below.

Above are some examples of grafting pitfalls resulting from a mismatch of rootstock and scion wood. The top left shows a 10-year old underdeveloped tree impeded by grafting, as shown close up on the top right.

Air Layering

This is an excellent way to asexually propagate a number of tropical and sub-tropical plants. It is a fast and easy way to create a new plant and one of the best ways to create plumeria trees of any diameter and heights up to seven feet. The best time to begin air layering for most cultivars is after the tree comes out of dormancy or in the early springtime. In Hawai'i, a good time to air layer species types is between August and November. Begin by cutting a one-inch girdle into the cortex all the way to the wood, and cover the girdle with moist sphagnum moss sprinkled with a rooting hormone. Wrap the moss with aluminum foil or three-mil polyethylene, folded side down, and tie the ends with wire so that moisture cannot enter. Clear poly makes it easier to see when the roots have developed. Roots will develop in eight to twelve weeks depending on the diameter of the wood and the cultivar. The rooted air layer can be cut from the mother tree and planted directly into a potting mix or the ground. This method has been proven to work on all species and cultivars at any maturity.

Steps in Making an Air Layer

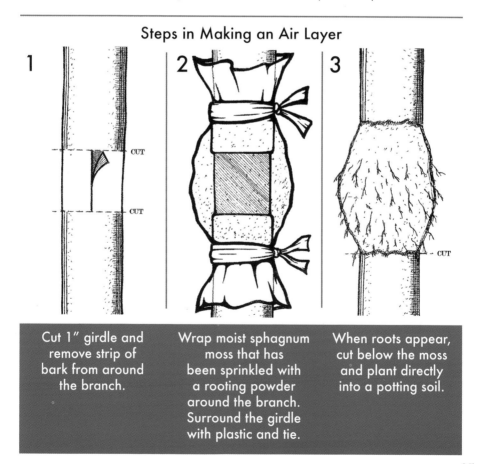

1 Cut 1″ girdle and remove strip of bark from around the branch.

2 Wrap moist sphagnum moss that has been sprinkled with a rooting powder around the branch. Surround the girdle with plastic and tie.

3 When roots appear, cut below the moss and plant directly into a potting soil.

Before you prepare a cutting to root in a container, you should first decide on what type of rooting mix you are going to use. The best rooting compounds are perlite, sand, pumice, cinder, or any loose textured medium used alone or combined with a little peat moss, which permits rapid drainage. A cactus mix which provides excellent porosity is another very popular choice for growers. Many growers also custom mix their components. Rapid draining is essential to faster root development. Coarse inorganic particles offer the fastest drainage. General potting mixes are often very dense and retain too much water for rooting. If used, they should be amended with perlite, Big R, cinder, sand, or other

Two-month old six-inch-rooted cutting in coarse peat moss and cinder

coarse compound. If you prefer to formulate an organic rather than an artificial soil mix, you should experiment or have the soil mix tested before embarking on any serious plant growing.

For rooted cuttings, there are many commercial potting mix formulas that can be used straight from the bag or amended by the gardener. Handmade mixes for rooted cuttings consist of equal parts of peat moss and perlite, or you can add soil or garden loam mixed with some bone meal and a time-release fertilizer. New roots often decay and the plant will die if too much organic material is added. Avoid food wastes and grass clippings.

Gang-planted seedlings ready to be divided

Three-inch tips rooting in perlite

If your soil is of good fertility and texture, it can be added to the potting mix. You can also root your cutting directly into a well dug hole in your yard by fortifying the soil with some cinders, sand, and/or nut shells, foam particles, or perlite. Regardless of the medium you choose to root your cutting in, it is essential to dig a big hole with good porosity that will permit optimum drainage with no standing water. Your tree will thrive as it develops a root system.

Cuttings rooting on top of the ground

Gang rooted two foot cuttings will make a massive flower display

Two months of rooting in peatmoss and cinder

Plumerias are happy to provide you with years of happiness without any special science or secrets. Just follow the suggestions provided below.

Watering

Bud set and flowering are determined by watering, daylight hours, temperature, soil mix, and fertilizer. These conditions should be balanced to provide the optimum results.

The key to watering is not to overwater. Overwatering prevents new fine rootlets from developing, reduces growth and metabolism, and causes buds to drop prematurely or even kills the plant. The better your drainage, the more often you can water and not worry about overwatering. It cannot be stressed enough how important rapid drainage is for plumeria in a pot or in the ground.

It is interesting to note that older established trees have a high tolerance for too much water. Plumerias are known to grow in some of the wettest subtropical and tropical areas in the world including along riverbanks and sandy shores of Hawai'i, where they are next to a continuous water table. Depending on the climate, sunlight, temperature, size of tree, and potting container, watering once a day is normally satisfactory. Water should be provided before the hottest time of the day. The leaves will begin to droop if the plant needs extra watering. Wa-

Overwatering and over-feeding leads to branches collapsing

tering methods include drip, automatic misting, water can, or hose. Any method used can be balanced with rainfall.

There are watering meters that can be purchased to determine the moisture content of the soil, but they are rarely used as water requirements constantly change. Water pH for natural water is 7.0 and, unless amendments to your water have been extreme, there should be no problems with using your existing water source. Municipal treatment of water supplies with chlorine is not sufficient enough to cause a plumeria plant injury.

If meters are used, testing should be done on established plants based on light, temperature, soil mix, and plant maturity. Meters can also be used to determine the soil pH, salinity, and fertilizer analysis. As your plant grows and matures, these requirements are subject to change.

Light and Temperature

Botany classes teach us that plants must have a proper balance of light, water, and nutrition to develop complete photosynthesis. With few exceptions, plumerias thrive in high light, not necessarily high temperatures. The epidermis (bark) of some plumerias is known to sunburn until the foliage grows out to help shade it. This is one good reason not to prune trees during the summer months. Sunburning is manifested as a crusty-bleached tanning on the bark and happens on newly exposed epidermis that is more sensitive to higher sunlight. Move the sensitive tree to a lower lighted area or turn the plant so that the foliage protects it from the direct sunrays, or spray the sunburned side with a pruning tar which will act as a protectant.

Left: Sunburned branches; right: pruning tar to prevent sunburning on sensitive cutting

Volunteer Pudica seedling with green foliage

Plumeria rubra cultivars will begin to flower as early as January in Hawai'i and as early as March in California, Texas, and Florida. Selected cultivars will bloom for six to seven months in Hawai'i and some states on the mainland depending on the daylight, temperature, and cultural practice. Trees that bloom early in Hawai'i are Tillie Hughes, Celadine, Donald Angus, Kauka Wilder, Indian Ivory, Tomlinson, Erma Bryan, Samoan Fluff, King Kalakaua, and Sherman. Trees that continually bloom throughout the year in Hawai'i with brief periods of rest are Celadine, Vera Cruz Rose and Penang Peach, which are commercially produced for lei flowers in Hawai'i.

Singapore obtusa varieties will bloom throughout the year, and will shed some leaves without going completely bare like the deciduous types. Longer days and more sunlight generate a longer flowering season. Altitudes over one thousand feet, above-average rainfall, and reduced sunlight all lead to reduced flowering. Areas that are closer to the equator will continue to flower up to seven months or longer. When temperatures go over a hundred degrees, it helps to mist the plant occasionally during the day or move it to a lower light area. Some varieties like Pudica (Bridal Bouquet) can be grown indoors with reflecting window light by using electrical grow lights. Blooming plumerias can be rotated from outdoor light to indoor window light, still retaining their flowers for short periods of time.

Plumeria flowers display higher color saturation if they are growing under soft cloud covers in very humid areas like Singapore and other Southeast Asian countries. A plumeria inflorescence in Hilo, Hawai'i, is deeper in color saturation than the same flowers from a tree growing in Kona, Hawai'i, due to the higher humidity and diffused natural light in Hilo.

Fertilizer

The three most important nutrients for plants are nitrogen for green and healthy leaves, phosphorous for healthy flowers and roots, and potash for overall vigor and disease resistance. The label NPK is used on all fertilizer containers to define fertilizer content. The higher the NPK total, the more costly the fertilizer.

In discussing the topic of fertilizer it is important to distinguish the difference between trees that are growing in the ground from those in pots. While each has different requirements, one of the best things about plumerias and fertilizers is they will tolerate any type of fertilizer, but they have been proven to produce more and bigger blooms using a high phosphorous content such as a 10–30–10 or similar ratio, depending on the manufacturer. For tropical and subtropical flowering trees and plants growing in the soil, many commercial growers use NPK 20–20–20 granular, liquid feed, or both. For potted plants, a recommended blend is Osmocote 14–14–14, a time-release fertilizer that lasts for approximately three months. Osmocote 18–6–12, also a time-release fertilizer, will work for about six months and is used by plumeria growers seeking a faster than normal growth rate. Time-release fertilizers can be top dressed in the pot or mixed directly into the potting soil. Other supplements such as bone meal or super phosphate can be added to the planting mix in the same way. Adding a little liquid fish emulsion or a few grams of Epson salts every month during the growing season provides an extra kick. A good potting soil with balanced nutrients will reduce your need to add any trace elements. Commercial nurseries supplement with a balanced liquid feed spraying every ten to fourteen days for potted plants and/or fertilize by polytubing to each plant controlled by an adjustable injection system.

New fields with young cuttings rooted directly in the ground

For plants in the ground the amount of fertilizer that you use should be regulated depending on your soil pH and the size and maturity of your plants. An acid to neutral pH (5.0–6.0) is best for many plants including plumerias. For feeding ground stock, Dr. Criley recommends using 1 tablespoon of fertilizer (one ounce) per one inch of the tree trunk diameter and spreading it around the trunk to the ends of the extended branches. As the trees begin to show signs of steady growth, increase your application rate. During the growing season, whether it be ground stock or potted plants, it is best to fertilize at half the recommended label over a longer period of time, rather than periodic concentrated one-shot applications.

Evergreen seedling

Different stages of deciduous tree defoliation

It is not necessary to fertilize after the end of the growing season. Over-fertilization in plants will cause a nutrient imbalance and a salinity build up that can affect the general performance of your plant. High salinity can lead to stunted growth, leaf burn or browning around the edges, and sometimes defoliation. If this happens, it is best to leach the pot with running water for about ten minutes a day for several days until the salt concentration is reduced.

Plumeria have been found to grow well in reduced soil content including volcanic rock and sand. One exception is heavy clay soil. Clay soils can be amended for suitable growing conditions. You can have your soil tested by taking a sample to a horticulture lab or university for analysis. They will recommend the proper fertilizer (NPK) along with any minor or trace elements and amendments that will help you develop the ideal formula for growing your plumerias in heavy soils. As the tree matures in a container or in the ground, a well maintained plumeria will produce bigger, more beautiful flowers.

Ground limestone and sulfur are amendments that can be added to raise or lower the soil pH. They are also useful in a dense clay-type soil. If you are planting in the ground with a neutral pH of 6.5 to 7.2, it is safe to top dress with a balanced (NPK 16–16–16) or similar balanced granular fertilizer for the most economical results. Plumerias are very forgiving plants and will give a lot of latitude for error. These recommendations have been tested and will help you begin a sound fertilization program. Variables do exist and a soil test will always confirm any special requirements.

Winter Preparation

It is important to remember that plumerias begin to shut down for winter in some areas faster than others. The shortened daylight period signals the plant that it is time to drop its leaves. Trees throughout the world will respond differently to seasonal changes. These differences exist from country to country, state to state and city to city. Unless you live in a subtropical area, plumerias have trouble surviving outdoor winters where temperatures dip into the thirties. To avoid injury to the plumeria it is best to prepare for "wintering over" when the temperatures begin to drop into the forties. It has been noted that some cultivars acclimate to frost or freezing temperatures for a short period of time with some tip burn, but with no serious damage. No studies yet exist that indicate what cultivars or species tolerate cold temperatures the longest. To date, there is no proof of so called "cold hardy" plumeria trees that can survive a season of intermittent freezing temperatures. While some Plumerias have been observed growing at high altitudes in different countries where cool temperatures exist without freezing, flowering is minimal.

Otherwise, it is best to begin preparation to move your potted plants into the garage or basement at the first forecast of cold weather, usually starting in November. As the leaves start to fall off, now is the time to stop watering, fertilizing and using pesticides. If potted plants were planted in the ground, they will need to be pried loose and can be stored as is or removed from the pots. Roots that have pushed through the drainage holes of the pot may be cut off. Some collectors conserve space by hanging the plants from indoor clotheslines and rafters or stacking them on shelves inside their garage or protected area. The plants are dormant so there is no need to worry about injuring the root system. If some leaves remain, just clip them off at the petiole (the stalk of the leaf) or let them fall off naturally.

If you decide to chance the winter season, bandage the tree with rags, frost blankets and water resistant tape or twine just as you would protect water pipes. Smudge pots, fans or other forms of heating can also be used effectively. Portable greenhouses will provide winter protection. To prevent an electric shock, strict precautions must be adhered to when working with electricity or hire an electrician to perform tasks that require a professional.

Plants in the ground may by dug out and root balled (with or without dirt) in fabric or newspapers, put in boxes, plastic bags (do not close bags), and pots or laid on cardboard old blankets or canvas. Do not put the plants on bare cement for storage. If root pruning is necessary, anytime is acceptable while the plants are dormant. The roots will stay dormant until springtime arrives. Some collectors tell me that they bring their potted bare bone trees inside the house from their garage around Christmas time, put them in a vase and, decorate them for use as a holiday ornament.

Growers can tell when spring is approaching because the tips of the mildly wrinkled and dehydrated plants begin to get a shiny look. They are saying that they have rested long enough and are ready to stretch their limbs for another

sp pudica in background displaying mass flowering during the winter months in Hawai'i

season. If the threat of frost is over and temperatures start approaching the fifties it is time to move them back outside, replant into larger pots, if necessary, and give them a light drink of water. In a few days, and as the weather temperatures begin to increase to around the high fifties to sixties, you will notice leaves starting to emerge from the tips of your plants. They are now ready to receive a little water and gradually their appetite will increase for more water and fertilizer as they renew their growth. Before you know it, you will be enamored by the colorful blooms and fragrances for another spring, summer and post summer season.

Branch Pruning

Pruning is done for several reasons: to remove diseased limbs, to thin out existing branches, to improve a tree's form, to remove broken or storm damaged branches, or just to propagate cuttings.

Pruning is best done during the early spring when the tree is still in dormancy and you can see the branches that are normally camouflaged by foliage. The correct way to prune is to cut (with pruning shears, razor saw, or chainsaw, depending on the size of the branches) flush to the branch collar to prevent any further branches from growing. If you desire more branching, leave four-to-six inch branch stubs coming from the main branch that is trimmed. New branches will develop from the remaining leaf node stubs. Information on branch induc-

Incorrect pruning technique   Correct pruning technique

ers using lanolin paste can be obtained by reading Plumeria PotPourri, Reasearch Bulletin – Vol. 1, No.2 by Dr. Richard Criley found on the PSA web site. To help prevent insect invasion, seal the pruning cut with a pruning compound or a water base paint after waiting a couple days for the cut to heal. Otherwise, the tree cuts should harden and callous over with no decay if left alone.

Proper pruning to induce branching

Pruning close to the branch collar to prevent new branches

Root Pruning

After several years, a potted plant begins to lose its quality blooms and growth, and needs to be rejuvenated. There can be several reasons for this: good cultural methods have not been practiced, the plant medium has been depleted and the roots are choking, or a seedling taproot has strangled normal root growth. Remove the root ball from the pot by either tapping the bottom with a sharp blow or cut the plastic pot open, and trim the roots back several inches depending on the size of the root ball. If a dominant root is twisted in the pot, it is a taproot and should be removed. Surgery to remove a taproot can be done anytime of the year. The plumeria will be invigorated and begin to spread new fibrous roots adding new vitality for another season of growth and flowering. Replant in a new enriched medium. Plumerias do not mind root pruning and any shock will be minimal and temporary.

Two-year-old OP seedling with first flowers

Mulching

Mulching is recommended to conserve water, implement weed control, or develop aeration in your soil. With most trees, and especially plumerias, it is most effective to use woody tree trimming mixes, not grass or chemically treated wood chips that have been treated to resist pests and may be toxic to growing plants. Spread the mulch evenly over the ground without touching the trunk of the tree. If you permit the mulch to touch the tree, you increase the chances of harboring foreign pathogens that will invade the trunk. A word of caution: some mulches will promote weed growth and pine needles will deplete the nitrogen from the soil. Preemergence chemicals that prevent weed seeds from germinating are Ronstar and Snapshot. Snapshot granular can be used safely in potted plumerias without causing injury to the plants. The residual effect will reduce weed maintenance for about three months. Remember that good weed control will help protect your plumerias from weed pests looking for more desirable plants.

Off-season and rust-resistant San Germain plumeria

PESTS AND WEED CONTROL

When discussing plant pests, it is important to be able to distinguish between plant growth problems caused by pathological abnormalities such as a virus or fungus, and those created by physiological disorders caused by plant culture and environment. If you are not sure, it is best to consult with an agricultural authority who can identify your problem

Plumerias have the tolerance to resist most common pests that threaten other ornamental trees. Insects do not find the rubbery latex substance in the tree very tasteful. Plumerias can play host to certain pests in selected localities. For example, a borer (*Lagocheirus obsoletus*) that occasionally attacks stressed or wounded plumeria branches in Hawai'i is noted to be uncommon in other parts of the United States and tropical America.

Conversely, black tip (sooty mold), a powdery fungus saprophyte that covers the leaves and branches, is rarely a problem in Hawai'i. It grows on sugary excretions from white flies, scales, and aphids harboring under the leaves. This

Typical signs of borer attack

phenomenon is often associated with microclimates of high humidity, excessive moisture and ants feeding off insects on the plumeria structure. This condition can be brought on by a weakened root system, which leaves the plant in a feeble state and manifests itself in the tip. Daconil™ (a systemic fungicide) sprayed morning and night to the branch tips or a pesticide root drench often clears up this problem and other related problems such as root mealy bugs or ants. The best way to drench a plant is to fill a five-gallon bucket with an insecticide mixture and submerge the potted root ball in the mixture for about five minutes. Another way is to fill a watering vessel with an insecticide mixture and pour it directly into your pot. For this treatment to create a fix and be most effective, it is best not to water for two days afterwards so as to not dilute your application.

Clockwise from top: Larva of the long-horned beetle, Lagocheirus obsoletus; Long-horned beetle; rescuing the cutting; larva extracted for study

Papaya mealybug (*Paracoccus marginatus*) is a recent problem that attacks not only papaya trees, but plumeria leaves and tips. It was discovered in Florida and is believed to have originated in Mexico or Central America. the pest can be observed as clusters of cotton-like masses on leaves and growing points that feed on the sap of the plant. The best management is to catch it early. Remove any infected leaves and apply Permethrin (Pounce™) or Malathion. Marathon™ or Merit™ used at the first signs is an effective insecticide control. Lady beetles (bugs) are known to be an effective predator. Once this pest becomes established, it is very difficult to control.

A fungal disease that presents itself as small purple spots on the leaves is called anthracnose, a leaf spot disease. It is also found on some vegetable and ornamental crops. High humidity and excessive moisture can contribute to the problem. The infected leaves should be removed and buried or put in a plastic bag for disposal. Commercial growers control anthracnose with Zyban, a basic contact systemic fungicide or Bayer Lawn Fungicide™. The recommended interval between treatments will vary depending on the severity and rate of application. Most fungi spread from plant to plant through touch, water, wind, and insects.

A widespread fungus in Hawai'i and many parts of the mainland where plumerias are grown is rust (*Coleosporium domingense*). Rust looks unsightly but does not retard the flowers or growth habit like sucking pests do. Honeybees are attracted to it by the thousands every year. Rust can be hand controlled by removing the older lower leaves from the branches. If you are a purist, use an insecticidal soap or even dishwashing detergents for ecological control, prevention, and treatment of infestations. Depending on the strength of the soapy mix, hose down the leaves after the spray has made contact with the bottom and top of the leaves for at least one hour. An insecticidal oil mixture can also increase control.

Bee collecting rust

Many plumeria evergreens and species including some cultivars are more resistant to rust then others. Singapore white and yellow (Mele Pa Bowman) and some evergreen dwarf varieties are among those most resistant. Species that have been observed to be rust free are: *pudica, caracasana, tuberculata, stenopetala, sericifolia, stenophylla (filifolia), alba,* and *bahamensis.* General granular systemic fungicides, such as Bayleton™, wettable sulfur, copper sulfate (liquid and dust), and water-soluble lawn and ornamental fungicides work best in controlling rust. Fungicides may be applied at anytime during the growing period, preferably at first notice or before the fungus becomes prevalent. Depending on the manufacturer, subsequent applications may be required to gain full control of the fungus.

Powdery mildew, another fungus, does very little damage to the overall plant health, but is unsightly. It can be controlled with Bayleton™, Funiginex™, Cleary's 3336 Domain™, or Banner™. It works best if used with a spreader-sticker added to the spray. Good air circulation around the plants will help prevent powdery mildew.

Insect pests are continually looking for a susceptible host plant to feed their appetites. One such pest is spider mite. Once the exact species of spider mite is identified, it can be controlled with an appropriate miticide. The two-spotted spider mite (tetranychus urtiacae) thrives during the summer season where the conditions are hot and dry. Many trees, shrubs and flowers play host to this mite. These sucking insects are first noticed under the leaves.

Tree joint destruction by the borer

Possible thrip effect

The Six-Spotted Mite (Eotetranychus sexmaculatus) is a problem for plumeria in the Sacramento Valley and Southern California. It can damage forming bloom clusters, cause premature leaf drop and decrease plant vigor. Deformed leaves during the first growth of the season may indicate the presence of this mite. A light horticultural oil spray (1% solution) can be used or insecticidal soaps can be used as an alternative for control. Repeat applications may be necessary.

Controlling these pests can be achieved by using Avid™, an emulsifiable concentrate, or Mavrik™, a broad spectrum pyrethroid. Horticulture oils or a soapy detergent mixture are often the best choice for homeowners if practiced on a regular schedule.

Another sucking insect is the thrip. While thrips do not cause damage to plumeria leaves, they will attack the inside of the flower corolla, preventing flowers from opening naturally, often leading to distortion, color fusion, spotting, and deformed flower buds. Some flowers have a hairy corolla opening that naturally helps to prevent insects from entering. Thrips are known to spread the necrotic spot virus in impatiens but we have no evidence to suggest plumerias are a host plant for this pest. Thrips are most prevalent during the warmer summer months and can be controlled with an appropriate insecticide or insecticidal soap.

Mealy bugs are small wax covered pests that suck the juices from a plant and secrete a honedew like substance attracting ants and other visitors to the tips of the plant. These bugs can best be controlled with a systemic insecticide that includes mite control or with persistent biological applications. They are also known to attack the roots of plumeria, both in pots and in the ground which will stunt normal plant growth. These root infestations are best controlled by using a dip or drench of an appropriate insecticide. This procedure also works for ant control.

Plumerias can play host to scale pests, which attach their sucking bodies to the bark causing tiny whitish round bumps. These include Oystershell scale or mining scale (*Howardia biclavis*), magnolia white scale (*Pseudaulacaspis cockerelli*), *Concharpis angraeci* scale, herculeana scale (*Clavaspis herculeana*), and armored scales (*Diaspididae*). Although there are several pesticides labeled for scale control, insecticidal soaps and oils (Neem and Volck™) are effective if you spray the plants every seven to ten days until controlled. A systemic insecticide will usually clear up the problem. For any pesticide spray to be effective, it must make contact with the top and bottom of the leaves. Adding a surfactant (sticker) to your insecticide and pesticide mix will deliver a more effective and uniform spray. A few drops of dishwashing detergent can be substituted for a surfactant. Visible scales on the branches (if only a few) can be removed by hand plucking or by rubbing the branches with a wet rough cloth with soap and water. If they don't pop off easily, then the bump is generally lichens, a normal part of the plant.

White fly (*Aleurodicus disperses*), which used to be the dominant pest in Hawai'i until the rust fungus overtook it, is active in all parts of the world. It is difficult to manage once it gains control of the plant. Whiteflies excrete honeydew, which lures ants and other nuisance insects onto the host plant and can lead to Black Sooty mold. Several controls that include White Fly Traps, Permethrin dust or insecticide, and Orthene (systemic insecticide) foggers are effective if used regularly. Spraying insecticidal soap, which is 100% safe even when used for vegetables or fruit trees, can also be effective.

Stem rot (a softening of the stem tissue), occurs when the cutting is being rooted. This usually can be attributed to over watering and premature applications of fertilizer (before the roots have developed). This same problem can occur with rooted plants. Once stem rot starts, the plant stem turns brown, weakens, and dies. It is best to recognize the problem at the beginning for the fungicides to be most effective. Attempting to recut and save the tip is often futile, even if a graft is attempted as a last salvation.

It should be briefly mentioned that, when using any granular or liquid herbicides for weed control, do not permit contact with the tree trunk, leaves, or flowers. Herbicides can seriously debilitate a plant or even kill it. If a post emergence herbicide is used, avoid making contact with the flowers. Herbicidal damage can sometimes be confused with a pest problem until testing is done, and is usually associated with careless application techniques or wind drift. The visual effect is a shriveling and narrowing of the leaves.

Many chemicals are restricted for use only by certified licensed commercial nurserymen, farmers, and pest control operators. There are, however, a multitude of chemical controls that can be purchased over-the-counter that will help with your problems. Always follow the brand name instructions on the labels when using any chemical. Consult with garden center specialists, a local farm extension agent, or a college or university in your area that has a plant disease clinic for testing plant irregularities. Labs can do an analysis that can provide insect identification, soil, plant tissue, and water and nutrient solutions information. While there are other pests that are attracted to plumerias, this information has addressed the most frequent visitors.

Ideally, if maintaining only a few plants, an ecological solution, such as using an insecticidal soap and water or hand management, is preferred over pesticides. Good cultural habits will promote good healthy plants and reduce the need for pesticides of any kind. Plumeria lovers are fortunate to be able to enjoy a plant with so few physiological or biological problems.

It must be remembered that plants most vulnerable to pest invasions are those that are neglected or stressed. As the blooming season comes to an end and the leaves begin to drop, it is recommended not to use any further pesticides.

Signs of herbicide damage to leaves in the foreground from wind drift

Mutations and Viruses

Many plumerias with pigmented flowers that look splotchy, variegated, spotty, or disfigured are not viruses, but a form of mutation (chimeras). This type of mutation can affect a few flowers of an inflorescence or a segment of only one flower. A mutation (sport) is just a change in the DNA of a plant and the change can be good or bad. Fasciated tree branching and form can also be affected. They are created by gene and chromosomal changes within the cells that can be heritable and rarely leads to permanent changes.

Above shows chimera effects on flowers. left: Moragne #23 chimera; right: Moragne #23 with normal petals

Examples of fasciated branching, peduncle development, seed set and inflorescence

Exceptions can be found as with the Florida, pink-fleshed grapefruits, which were discovered on a single branch of a tree in a grove of thousands of trees producing white-fleshed grapefruits, an obvious mutation that held true to the parent when propagated. These so-called mutations occur in plumerias, roses, poinsettias, orchids, and anthuriums, to name a few. Other plant irregularities can occur from chromosomal changes that lead to diploid-tetraploid mutations. This will be noticed in a plumeria with leaves as thick and stiff as a shoe heel. Normally, a tetraploid will have sixty-four chromosomes per cell whereas a diploid will have thirty-two.

Variegated foliage is an example of a chimera in which the leaf pattern is mottled white or mosaic. When this occurs in plants, parts of the leaf tissue lack the capacity to produce chlorophyll. Most often variegated foliage will revert back to all green as the plant matures. As of this writing, there is no plumeria tree where all the flowers display total variegation or a mature tree with totally variegated foliage.

Some kinds of leaf variegation are not chimeras, but are due to virus diseases, an important distinction. In a viral infection, the leaves turn anemic, yellow, and suspend downward. When a virus does occur, it is often in palms, bananas, and ornamental foliage, and seldom in plumeria flowers. Such viruses will not cross-over to plumeria. Once detected, infected plants of any kind must be burned or buried to arrest contamination of surrounding plants.

Because chimera irregularities are genetic and propagation from plants by seeds, cuttings, grafting, or cloning are not predictable, heritable changes are not guaranteed. Chimeras will not transfer to surrounding trees or plants of the same or different species. Chimera flower and foliage irregularities can appear on any cultivar of the same species spontaneously one time or several times seasonally or annually. Propagation sexually or asexually is no guarantee of repetition.

Bali Whirl ten petal mutation that holds true

Opus #1 JL seedling Ruffles mutation holding true

Tidal Wave JL seedling mutation holding true

Seven Petal anomaly

Anomalies

Anomalies are irregularities that sometimes occur in plumeria flowers. Normally a plumeria flower has five petals, but sometimes the same tree will surprise you with six, seven, or even eight petals. An exception is the Bali Whirl flower with ten petals, or "ruffles" flower types. A tree with a mutation will not necessarily be able to replicate itself from seeds of the same tree. Out of over one hundred Bali Whirl plumeria seedlings, not one has produced a ten-petal flower. (See ten-petal Bali Whirl pod, page 60.) Bali Whirl is a mutant (sport) from Celadine and its seedlings have reverted back to five-petaled yellow flowers.

In some cases, deformed, spotted, or mosaic flowers are not a mutant, fungus, virus, or anomaly, but are the result of mites or thrips that have penetrated the flower corolla tube causing it to abort short of developing into a normal flower. The petals of the flower fuse together and cause a color abnormality. Other complications can occur when the flowers are affected by fungi in the petals. Many of these abnormalities are here today and gone tomorrow. They do not repeat themselves. Anomalies can sometimes be confused with the effect of excessive heat or high humidity which retards the floret from fully opening on schedule. This could result in color fusion or splashing pigmentation in the flower petals. Flowers that display this appearance are sometimes mistaken as

Pink Twirl anomaly

Anomaly

Bill Moragne four petal anomaly

viruses. Trees should be observed for a minimum of three years to determine if the tree flower irregularity will hold true, and whether the flower abnormality is caused by mites, fungus, or is a chimera or an anomaly. Plant lab testing will further help to confirm suspicious discoveries.

Part of my plumeria collection is made up of all kinds of genetic and chromosomal irregularities—weird growth habit, shapely leaves, funky flowers—and they are referred to as my circus plants. One plant in particular that has caught my attention is a multi-fasciated plumeria that is one of the most intriguing and rare plants in the collection.

Chemical Experimentation

Dr. Richard Criley, University of Hawai'i professor and researcher in horticulture, field tested Celadine, common yellow, with a spray mixture of Ethephon, 800 parts per million (ppm) of active ingredient. This chemical, which is used as a growth regulator for fruit and nut trees, is being tested to see if it will stimulate off-season flower production in plumerias. It appears that, if applied in early September, it is possible to increase flowering about ten to twenty-five percent through December. If these field tests continue to be successful, the flowering season for some plumerias could be extended, and it could be used for inducing flowering in potted plumeria plants. Other tests using the chemical Cytokinin that induces leaf retention and branching were reported by Kwon and Criley in *Horticulturist Digest, volume 93.*

Another chemical that needs further testing is Gibberellic acid. It is used primarily as a plant growth regulator and has been experimented with on seeds. Research is needed as to the long-term effect on trees and seeds from chemical applications.

Ethephon treatment to promote winter flowering on Celadine

Wind-toppled Pudica

Natural disasters

Natural disasters are unexpected. If severe, damage from wind, rain, or floods may topple your plumeria tree. If you don't mind the looks, just leave the tree on its side and new branches will emerge upright from the main trunk and the existing branches will return upward. Bill Moragne called this to my attention when his entire collection of trees was whipped to the ground by Hurricane Iwa in November 1982. He just left them where they fell. The trees grew back to be even more prolific. You can also cut the plumeria trees into logs (up to seven inches in diameter by four feet long) and plant them in the ground. To reduce transpiration, cut some of the tips off on a few branches. The tip cuts can be propagated and the tree logs will root just like a cutting, and the cut branch tips on the fallen tree will divide into new shoots as the tree recovers.

Princess Victoria suspected chimera

Top: P. *obtusa* ssp. *sericifolia* var. *tuberculata* with typical knob-like branches; Below: P. *obtusa* ssp. *sericifolia* var. *tuberculata* with rust free foliage

MORAGNE HYBRIDIZATION AND CROSS-POLLINATION

The *American Horticulturist*, April 1991, featured an article on Bill Moragne (1905–1983) and his hybridization techniques. This article, coauthored by Dr. Richard Criley and Jim Little, answers many questions on how to cross-pollinate plumerias.

After lengthy attempts at different methods of pollinating plumerias, Moragne discovered that by carefully placing his chosen pollen grains to the base

Cross Section of a Plumeria Flower

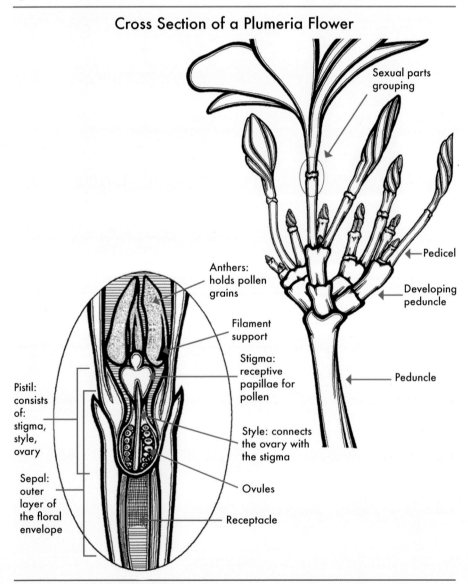

Sexual parts grouping

Pedicel

Developing peduncle

Peduncle

Anthers: holds pollen grains

Filament support

Stigma: receptive papillae for pollen

Style: connects the ovary with the stigma

Ovules

Receptacle

Pistil: consists of: stigma, style, ovary

Sepal: outer layer of the floral envelope

of the pistil through an incision cut into the side of the flower petals (corolla) and using plastic tape to secure the incision from insects, pollination actually took place. Fourteen days later, he noticed that a seed pod was beginning to develop at the base of the flower (pedicel). His first and only cross was made with Scott Pratt (formerly Koloa Red) as the pollinator and Daisy Wilcox (formerly Grove Farm) as the receptor. Nine months later, the seeds were harvested and 283 seedlings were germinated. He kept his best thirty-five seedlings, naming

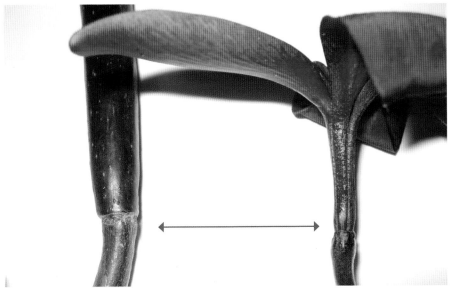

Moragne incision point for pollen application

Mary Pukui pure yellow semi-evergreen tree

eleven hybrids for his family members. The rest were numbered and given to friends and donated for landscaping. Moragne is the first person to document his crosses and to make a major breakthrough in cross-pollination. For this new generation of plumeria flowers, we will forever be indebted. Moragne proved that deliberate cross-pollination is possible and provided a germplasm for future experimentation.

We know that plumerias are bisexual (contain reproductive organ parts of both male and female). Natural or open pollination of plumerias can occur in many ways. The period during which a flower can participate in the pollination process is called anthesis. Male anthesis is when the anthers split open and release their pollen; female anthesis is when the stigmas are spread open, fresh and glistening, and slightly sticky. Male and female anthesis in plumerias can occur at anytime of the day, but not necessarily in the same flower on the same day.

While there is no scientific evidence to conclude that thrips are a source of pollination, they have been observed here and on the mainland penetrating inside the plumeria flowers. The same observations have been made in other parts of the country, where moths, butterflies, and other insects are assumed to pollinate. Wind, rain, water, dirt, and canned air may all contribute to self-pollination. Whether legend or myth, it has been said, in rare circumstances, even birds, bees, or butterflies (known to see in color) can set them off. Every natural pollinator is attracted to some feature of a flower: shape, size, color, fragrance, or another predator on the flower. Could this be one of the reasons that it appears that many plumeria trees with the most fragrance produce the most seed pods? Examples of such trees are Samoan fluff and Vera Cruz Rose. Upon dissecting plumeria flowers, I personally have observed many types of insect parts, including a bee tongue and the leg of a

Left: Curious invader; Right: Mary Helen Eggenberger after cross-pollination

praying mantis. The pest that naturally creates self-pollination must find its way through the small corolla (orifice) of the flower to the pollen on the anther.

No matter what your hand pollination technique is you must sterilize your applicator and emasculate (remove the existing pollen from the selected female flower). This will increase your chances of not mixing the new pollen with the existing pollen. After you have made your attempted pollination, you must cover the flower or flowers with a piece of nylon mesh or use plastic tape or a straw securing the end so that the flowers are protected from outside pest intruders.

Moragne did his experiments in the morning. While this time is used by hybridizers of many types of flowers, I have found that you can actually self-pollinate plumerias at any time of the day or evenings. This observation was made watching different insects visiting the flowers at various times. I noticed that during this time the receptive collar of the flower is enlarged and the stigma has a glutinous-like quality that is prime for pollination. Further, observing what trees set seed helps to identify a receptive parent.

Most efforts to hybridize plumerias use the artificial proboscis method emulating moths' method of natural pollination. These hybridizers may use wire, bamboo splints, toothpicks, horsehair, palm fiber, air, water, and other probes hoping for a deliberate cross. Regardless of these attempts, we must realize that the existing pollen on the mother plant must be completely removed to use the Moragne technique mentioned above before we can be assured that a successful cross-pollination can take place. Of the two methods mentioned above, I have found that Moragne's proven method is the best way to reduce the chances of a self-pollinated seedling.

Left: Straws for incision protection against invaders; Right: Three new seed follicles from a cross

We can learn from experiments by orchid fanciers who have dedicated their lives to cross-pollination. Most orchid growers, who make thousands of crosses yearly, keep less than 1 percent for further breeding. In addition, it is not uncommon for a cattleya hybridizer to wait seven years on average to see his/her first flower.

Deliberate crosses that are made with exceptional parents can produce an exceptional offspring making the seeds extremely valuable. The seed pods must be correctly labeled with the male pollinator and the time and date the cross was made. Once success comes to you, you can start an effective breeding program, which can lead to new breakthroughs in color barriers, shapes, and fragrances. The hybridizer should protect the seeds from viewers who often do not realize the value of a rare cross and may help themselves to the seeds.

The success of deliberately cross-pollinating plumerias is based on four very important steps:

1. Study which flowers will successfully cross (naturally set seeds) and use these trees as your receptors. Two known trees that have proven themselves are JL Pink Pansy and Daisy Wilcox.
2. Determine the time of year (best month, week, day, or time of day) to make the actual cross. Many trees are not fertile twelve months of the year.
3. Follow Moragne's proven method of self-pollination.
4. Develop the patience and persistence to make it succeed.

Following these instructions and using deductive decision making will lead to success in plumeria hybridization.

Naming and Registration

An open-pollinated or deliberate cross that produces a flower may be given a proper name and be registered with the Plumeria Society of America. The PSA is the International Registration Authority for Plumeria Cultivars. Upon application approval, the given name will be officially registered and dated by the PSA and a Certificate of Registration awarded the registrant. As of 2004, there were over six-hundred registered cultivars with registrations increasing yearly. It is estimated that there are over fifteen-thousand unnamed seedlings being cultivated by hobbyists and collectors. These numbers will continue to multiply yearly. Information on registration can be found on the home page of the Plumeria Society of America using the URL: http://www.theplumeriasociety.org, or by writing to the PSA, PO Box 22791, Houston, Texas 77227–2791.

Clockwise from top: Dwarf Evergreen Singapore white; Dwarf Ornamental type; Dwarf pink glow seedlings

Dwarf Plumerias

Dwarf plants are highly desirable as they can be grown in limited space on a lānai or in the garden. When moved inside during the winter months, they take up less space. Some collectors specialize only in collecting dwarf plumerias. There are approximately sixty true dwarf plumeria types, some deciduous and some evergreen. JL Nursery & Farms is developing a Little Darlin' series where the plumerias are not over four feet high. Their forms are particularly interesting. Many are extremely compact while some look like a candelabra or are shaped like a cactus. In the very near future we will begin to see more assorted colors.

Dwarf compact Rainbow JL seedling

Pompom dwarf shrub–JL seedling

Sometimes trees are referred to as dwarfs when they are actually not. Seeds from a dwarf tree do not always produce a dwarf offspring. A button-size flower on a tree is not necessarily a dwarf tree. True dwarf trees are associated with distinct compactness. Some dwarf trees are more compact than others and classifications of dwarf types should be established. Dwarf trees can range from two to three, four to five, or six to seven feet. Anything over eight feet outgrows its dwarf state. This guideline could be developed for a ten year old plant. All dwarf trees eventually get big but their period of growth is slower and their branching compactness can vary. For example, some have very stringy branching and, while it may be called a dwarf, they have the same elongated branching as normal trees sometimes do. Also, seeds from evergreen dwarfs, just like evergreen Singapore obtusa types, do not always produce evergreen seedlings.

Rare and Favorite Cultivars

Since Bill Moragne's hybrid plumerias were first developed, countless new seedlings have originated from his trees with some of the most exciting color combinations and fragrances and flower sizes ever encountered. Color combinations have melted together to create a melting pot of rainbow hues, with fragrances sweeter than some expensive perfumes.

Propagation by tissue culture or meristem continues to be a challenge for plant propagation labs attempting to replicate selected cultivars. While success has been achieved, we must ask the question, is it economically feasible to tissue culture plumeria? Plumeria researchers and hobbyists continue to experiment with this amazing plant. We are still in the infancy of plumeria mania and, as the popularity of new seedlings sparks our anticipation and imagination, we will continue to see plant collectors join in this rewarding hobby.

Dwarf Cabbage Patch JL seedling

Black Ruby JL seedling extremely rare red flowers with black veining

Hawaiian coral JL seedling

Mango Blush

Lilac Clouds

Foxtrot JL seedling

Hawaiian Ribbon JL seedling

Bud Guillot's 'Wild Fire' aka Herman Shigemura

Rising Sun JL seedling

Eclipse JL seedling

Don Ho JL seedling

Purple Rain JL seedling

Hawaiian Fire JL seedling

Top left: Majestic JL seedling; Top right: Mr. Ambassador JL seedling;
Bottom: Haleakalā JL seedling

Olympic Torch JL seedling

Purple Rain JL seedling

Hawaiian Rose JL Seedling

Hawaiian Classic JL seedling

Pink Lady Petal-cupped JL seedling

Liliko'i JL seedling

Fantasia JL seedling

Tempest JL seedling

Sensation JL Seedling

Hawaiian Style Soft-colored JL seedling

Red Spice JL seedling with mottling

Don Ho JL seedling

Sparkler JL seedling

Genseng JL seedling

Magma JL seedling with red veins

Happy New Year JL seedling with ribbon-colored petals

Sumatran Tiger Eye Private collection

Spirit JL seedling

Plumeria Species

At this time, to comment on the true species, the number and names would be premature. As of this printing, taxonomists have not done a definitive study on plumeria species to give an accurate account of how many species actually exist. Part of the confusion can be traced to the many synonyms and subspecies of the same species. Seeds from the same species do not always replicate the exact species. It has been rumored that there are at least nineteen different species. Until we can develop a more comprehensive grouping for these trees, it is best not to speculate. According to Dr. Criley, "The only published monograph on plumeria was done by R. E. Woodson, Jr. in 1938—*Annals of the Missouri Botanical Garden 25:189–224.*" Only nine species are identified.

Top left: sp. alba with seedpod; Top right sp. stenopetala seedling was planted from a seedpod from sp. stenopetala, pictured below, illustrating that species seedlings do not always come true to the parent.

LANDSCAPING

Plumerias of all sizes adorn the landscapes of golf courses; hotels; city, county, state, and federal grounds; botanical gardens; universities; colleges; and homes. If there is limited space for planting, plumerias can be grouped in a confined area using dwarf or compact trees. A larger residential landscaping area would require a minimum of ten-foot spacing depending on the mix of plants in your design. Some people with minimum space plant their trees in jungle fashion with spacing five to ten feet apart. The trees will mingle with each other to create a beautiful setting of color and foliage.

Plumeria color choices for landscaping

Shaped plumeria tree surrounded by plants and palms

Dense plantings should be arranged by the growth and branching habit of the plumerias. The more vigorous trees in a cluster planting could be trimmed to make the plantings uniform. If you decide on dense plantings, you should be aware that this will lead to increased root competition, and reduced light and air circulation, which are all major factors contributing to tree stress.

Plumeria evergreen trees (singapore obtusa) for future landscaping projects

JL Hula Girl is one of many landscape plumeria color choices

Commercial plumeria growers generally space their trees, depending on the growth habit, anywhere from sixteen to twenty feet apart, anticipating at least a ten-year growth period.

Some trees are well suited for xeriscaping such as Singapore white and yellow (Mele Pa Bowman). Dwarf varieties, especially Dwarf Singapore pink (Petite Pink), and some species can survive intermittent dry periods. In addition, a few deciduous trees such as Kimo, Celadine, and Candy Stripe acclimate easily to xeriscaping. Many older trees with large root systems can go weeks or even months without water, but this will stress them and render them more susceptible to problems.

Vertical-growing pudica plumeria

Large trees can be transplanted for an immediate visual impact. Removing or transplanting large trees should be done by professionals. To be successful, preparation of a rare or valuable tree for removal is essential. Failure to plan ahead can result in damage or even death to the tree. Preparation, if done prop-

The steps in removal and relocating a plumeria tree

erly, begins by trenching a square or circle around the tree, three to five feet deep depending on the size of the root system, using a backhoe or ditch witch. A commercial tree spade can also be used to penetrate the root ball and lift the tree directly from the ground, but will often break the higher plumeria branches. Another method for lifting the tree is to use a boom crane. The boom cable and hook drops through the middle of the tree and hooks around the base of the trunk, minimizing any breakage.

The tree trunk is prepared by wrapping it with a collar of heavy fabric so that, when the lifting strap is wrapped around the trunk, it will not rip or injure the bark. If the bark is stripped from the trunk, the tree will usually die. The tree is then lifted by the boom truck to a flatbed truck and taken to its destination where a large hole twice the size of the root ball has been prepared. To minimize transpiration and breakage some tree movers wrap the root ball with heavy duty Saran™ Wrap to hold the ball intact when relocating. After the tree has been transplanted, it is back-filled with soil and watered immediately, supported with guy wires or poles if necessary, until the tree is secure in its new location. Plumeria trees of all ages transplant easily if done correctly and will continue growing, often without the leaves dropping or any shock to the tree. Large plumeria trees sell for $1,000 to $1,800 depending on the diameter of the trunk, the height, and the cultivar. Labor and equipment costs are additional and charged by the contractor depending on the agreement. Other methods of removal and transfer can be used as determined by the grower and buyer. Finally, the landscaper must decide between an evergreen or a deciduous tree.

"Hong Kong" obtusa surrounded by deciduous plumeria trees

MARKETING PLUMERIA FLOWERS

During and after World War II, plumerias were the chosen flower for lei. Most flowers were collected from home gardens and public places and sold to the local lei stands, airport, cruise liner docks, and Waikīkī. They remained popular until the 1970s when orchids from different parts of Southeast Asia were starting to be imported at a lower price. These continued to gain in popularity until the late 1990s when the locals and visitors to Hawaiʻi desired the beautiful fragrance of plumerias and their new colors. Plumerias are Hawaiʻi. As a result, the demand for plumerias reasserted itself and plumeria farms that once cut back on production

Flower picker and lei maker

From top left: Stringing lei for local sales; ceremonial lei; lei-making in Hawai'i; flower lei for sale ; grandson Dane Little working on the farm

began replanting to meet the increased demand for this exotic flower here and on the mainland. According to the Hawai'i Agricultural Statistic Service there were thirteen commercial plumeria farms reporting sales in 2004. There are several plumeria farms being developed in California and Florida that are specializing in lei flowers. Prices for fresh plumeria lei that were once $3.00 are now found selling for $7.00 to $15.00 depending on the lei flower color combinations and freshness. Two new popular flowers added to the plumeria lei selection are Vera Cruz Rose and Penang Peach. Vera Cruz Rose has the same true fragrance of a rose and is fast becoming as popular as common yellow for lei when it can be found. Plumeria leis on the mainland retail for $35 to $50.00 each. There are fifty to sixty flowers per lei and commercial growers charge pickers anywhere from $15.00 to $25.00 per five-gallon bucket depending on the time of year. The flower picking is mostly contracted out to vendors who harvest, string, and sell the lei.

State Foundation on Culture and the Arts   Art in Public Places

LARK GREY DIMOND-CATES
THE LEI MAKERS
Bronze, 1997

In Hawai'i, the lei, a garland of flowers, conveys respect, love, friendship and the continuity of life. It may be created from a wide variety of local flora and is offered on many occasions calling for a special token. The giving of leis as a floral tribute symbolizes the way in which Hawai'i welcomes its visitors and bids them safe journeys home. These lei makers are surrounded by bas-reliefs depicting the cultural and physical features of the various districts of the island of Hawai'i.

Lark Grey Dimond-Cates was born in Kansas City, Missouri in 1953 and has lived in the islands since 1984. She has had a continuing interest in the cultural traditions of Hawai'i as embodied in the contributions of island women.

State of Hawai'i   Department of Accounting and General Services

State Foundation of Culture and the Arts monument, *The Lei Makers,* at the Kona Airport

Penang Peach lei flowers ready for picking

Six-inch rooted cuttings in two-inch pots for local sales

105

COLLECTING PLUMERIAS

Collecting plumerias has turned into one of the fastest growing plant and gardening trends today. From the beginning of time, plant lovers have indulged themselves in their appetite for exquisite beauty. Valuable collections include some of the rarest plants selected from all types of tropical flora. As in any hobby, there are very serious collectors who delight in spending hundreds of dollars on the rarest named exotic cultivars in existence and the amateur who is content spending five dollars on common or no-name varieties. There is a wide range of pricing and, in most cases, a cheap buy equates to a cheap deal. Caveat emptor! Both types of collectors have a love for the plant and, regardless of economic status, a collector can always find an affordable plumeria that suits his/her tastes.

Durban Yellow

Each plumeria has its own uniqueness. There are customers who only collect a particular color and others who try to collect one of everything, which is difficult. Still others collect and concentrate on dwarf cultivars. The most serious

Thornton's Mardi Gras

Guillot Sunset

Dwarf deciduous yellow

Duvauchelle

Fruit Salad bouquet inflorescence

collectors look for atypical flowers, tree shapes, patterns, and sizes. Regardless of what type of collector, once you are hooked, it's hard to let go.

In recent years, people have traveled to subtropical countries in hopes of finding rare, undiscovered plumerias only to find that they simply do not exist. Most third world countries have only common pinks, yellows, and whites. Occasionally you may find something extraordinary, not on the public streets where most commons exist, but in someone's yard. This is particularly true in Hawai'i. People in Hawai'i love their plants and some of the best collections are growing in backyards. The cultivar Cerise, common yellow (both with many synonyms), and Singapore obtusa are the most common of all plumerias found in Southeast Asia, Mexico, and other subtropical regions of the world. These trees were early introductions and were the only ones available at the time to propagate for landscaping.

Landscapes are slowly changing as new cultivars are exported from Hawai'i and other parts of the world and more people grow exotic seeds from hybrid trees. Singapore Botanical Gardens, for example, has a fine collection of named hybrids. Many of the plumerias in Singapore originally came from Hawai'i. They have provided a germplasm for new cultivars. Dr. Criley provided Singapore Botanical Gardens with their first wave of plumerias in 1972. Jim Little donated new cultivars in later visits.

The diversity of local plant lovers in Hawai'i and existing resources has led to countless new generations of seedlings and created a reservoir of new plumerias. Countries that are collecting from this seedling base and creating new plumeria hybrids of their own include Singapore, Thailand, Indonesia, Australia, India, and, most recently, South Africa, China, and selected localities of Europe and the UK.

Something to be aware of as a result of the rage to collect new and different plumerias is that some vendors are making false claims here and abroad to capitalize on this growing popularity. Valuable collections have also experienced thefts ranging from "just a cutting" to whole trees. Additionally, agriculture theft and crime cost Hawai'i farm producers $11.4 million in 2004. This has prompted some farms and nurseries to discontinue hosting visitors. Total security to prevent theft and/or vandalism to equipment amounted to $7.4 million for all farms.

Puu Kahea

Paul Weissich

Dwarf Pink JL seedling

Madame Pele

111

Dwarf Evergreen Rich Criley Rainbow

JL Pink Pansy

Metallica JL seedling

Mr. Ambassador JL seedling—named for Emerson Willis

113

Henry "Apples" Dupree–prolific bloomer with intense fragrance

Taking plumerias home from Hawai'i is encouraged if proper procedures are followed. State and federal agricultural personnel work hard to prevent threatening foreign pathogens from entering and leaving Hawai'i. We must do all we can to be part of this prevention.

To prepare a cutting for inspection, strip the leaves and wash it with soap and water; take it either to the USDA at the Honolulu airport or the Hawai'i State Department of Agriculture, 1849 Auiki Street, Sand Island, Honolulu, Hawai'i, for inspection. To prevent any potential nematode contaminations, do not let the cutting make contact with any soil. Plants with soil will not be approved. Bring a box and tape to pack the cuttings. Once cleared, the inspectors will stamp the box with an agriculture stamp, and you can proceed to carry it on board with you or mail it. They will provide you with a Phytosanitary Certificate if required (principally for foreign countries) and any other instructions. If you do not have the time for this, the easiest way, of course, is to purchase a packaged cutting from a certified Nursery or a gift shop that carries certified plant products. This way, you are guaranteed a cutting that is free of pests and can be hand-carried or mailed directly to your home in any state. It is illegal to ship a plant through the mail or hand-carry it for travel without an agricultural stamp or documentation. For importing plants to Hawai'i from the U.S. mainland or internationally it is best to first check with the USDA to learn what the requirements are. Some plants are prohibited.

Entrance to Koko Head Crater

Little's Mosaic

Forty-year-old tree at Dean Conklin Plumeria Grove

In your home state look for nurseries and garden centers that not only sell plumeria, but can answer questions. Reliable sellers to buy from can be recommended by the PSA and other plumeria societies mentioned above. Other sources include plumeria web sites and, most recently, eBay. A proven and creditable eBay source for a large selection of plumeria cuttings and seeds is Hawai'i/Sandy, which lists popular favorites along with new yearly introductions.

116

Visitors to Hawai'i should see Koko Head Botanical Garden to enjoy plumeria trees that are over forty years old. The garden was acquired in 1957 and developed and operated by the city and county of Honolulu. In 1962, the former director of Foster Botanical Garden, Paul Weissich, was instrumental in the formulation and development of this two-acre piece of arid land into a showcase of plumerias collected over the years by Dean Conklin, Bill Moragne, Dr. Richard Criley, Donald Angus, and Jim Little. The garden was dedicated to Dean Conklin, the original volunteer who, for years, canvassed the streets of Honolulu and the neighboring islands looking for new plumerias to add to the collection. A word of caution: some accession tags that once properly identified the trees have either been switched or stolen and names of the various trees cannot be relied upon.

Above: Monument to Dean Conklin Plumeria Grove. Below: Dean Conklin plumeria

CL Allison cv., named for Clark Little's daughter

No criterion yet exists to judge the ideal characteristics of a plumeria flower or tree. All plumerias have their own personality and character and the variations that occur in plumerias makes each tree unique. With so many features to select from in a tree, most people make their choice based upon color, fragrance, or flower size. Some collectors claim to be able to identify cultivars by their fragrance only.

As the eye develops, the sophisticated collector will begin to notice the more subtle features of the flower, the different petal shapes, pattern rollback, and underside color variation. Notice the peduncle (the stalk that supports inflorescence) can grow pendulous or upright, as can the pedicel (the stalk to each individual flower in an inflorescence). You will see the number of florets in an inflorescence and the way in which the pedicel displays the showy flowers. Also note the broad differences in leaf structure shapes, shades, and texture, and see the variation in the tree bark, texture, and color; notice the growth habit, which can be lanky, stringy, bumpy, beefy, vertical, or horizontal.

Your choice of the most desirable plumeria should be what pleases your eyes and nose. Depending on one's level of taste and experience, one collector may pay hundreds of dollars for the most exotic varieties while others with less exposure will be happy with a $5.98 common cutting. Collecting can be more fun if one collects for quality cultivars rather than for quantities. One of the best things about collecting plumerias is the countless variety. New ways of cultivating plumerias for interiorscapes are being experimented with. There is still much to study and learn from these exciting plants and we are only on the threshold. We hope that soon DNA research will be able to map the most desirable characteristics of plumerias. We will then be able to clone them in a test tube and propagate only the very best trees and flowers. So far, tissue culture continues to be a challenge with plumerias. A few labs are showing signs of success. Dwarf Petite Pink has proven it can be cultured. Labs and individual experimenters continue to make progress. The economics will play a factor to see if tissue culture for plumerias can be profitable.

It is important to remember that each plumeria tree has its own biological and physiological composition just as people do. While some differences are minuscule, each plumeria tree is unique in some way. Further information can be found in the *The Handbook on Plumeria Culture* by Richard and Mary Helen Eggenberger, and other readings listed in the selected references of this book.

Finally, protecting the integrity of the use of proper plumeria names is the responsibility of all plumeria owners, whether seasoned veterans or newcomers. To learn more you can join your local amateur or professional nursery and farm association, garden club, or plumeria societies or take some extended day classes at your local college or university. Don't forget to visit the botanical gardens in your hometown and when you travel. You will be surprised at how many others share your plumeria passion and interest in plants.

Snowball JL Seedling

SELECTED REFERENCES

Armstrong, Helen, "Plants don't need so much phosphate!" Flower Tech, *Horticulture World*, Volume 8, No. 3, 2005.

Chinn, James T., Horace F. Clay, James L. Brewbaker, and Donald P. Watson, "Hawaiian Plumerias," *University of Hawai'i Circular 410*, November, 1965, reprint, June 1972.

Chinn, James T., and Richard A. Criley, "Plumeria Cultivars in Hawai'i," *University of Hawai'i Research Bulletin 158*, March 1982; reprint, February 1983. http://ctahr.hawaii.edu/cc/freepubs/pdf/OF-31.pdf

Criley, Richard A. 2004. Plumerias in Hawaii: http://www.ctahr.hawaii.edu/oc/freepubs/pdf/OF-31.pdf

Criley, Richard A., and Jim Little, "The Moragne Plumerias," *American Horticulturist*, April 1991.

Eggenberger, R. and M.H. Eggenberger, 2000. *The Handbook on Plumeria Culture*. 4th Ed. Tropical Plant Publishers, Cleveland, GA. 115p.

Essig, Frederick B. The Birds and Bees, *Florida Gardening*, Oct./Nov. 2004.

Green, Ted. *Orchids in Hawai'i*, Mutual Publishing, LLC. First Printing, 2005.

Harrington, H. D., and L.W. Durrell. How to Identify Plants. Chicago: Swallow Press Inc., 1957.

Hartmann, Hudson T. Plant propagation, seventh edition. Englewood Cliffs, New Jersey: Prentice-Hall Inc., 2002.

Henley, Richard W. Foliage Plant Problems, Florida Foliage Association, 1983.

Kwon, Eunoh and Richard A. Criley. Cytokinin and Ethephon Induce Greater Branching of Pruned Plumeria, Horticulture Digest; Hawaii-Cooperative Extension, No. 93, March 1991, p. 6-8.

Miller, Sean. 2006. The Six-Spotted Mite. www.plumeriatc.org.

Natural Agricultural Statistics Service, Department of Agriculture, Hawai'i Agricultural Theft and Vandalism Report, October 17, 2005.

Staples, George W., and Derral R. Herbst. *A Tropical Garden Flora*, Honolulu: Bishop Museum Press, 2005.

Thornton, E. and S.H. Thornton. 1978. "The Exotic Plumeria," *Plumeria Specialties*, Houston, TX.

UBC Botanical Garden and Centre for Plant Research. http://ubcbotanicalgarden.org.

GLOSSARY

Acuminate: Tapering to the apex, the sides more or less pinched in before reaching the tip. Compare acute.

Acute: Tapering to the apex with the sides straight or nearly so, usually less tapering than acuminate.

Anther: Male reproductive part, which produces the pollen in a flower. The pollen-bearing part of the flower stamen.

Anthesis: Period when the flower is open.

Apex: The tip portion of the leaf blade.

Asexual: Plant reproduction by a limb, branch or stem cutting, bulb, rhizome, or other vegetative plant part. Tissue culture is also an asexual method of propagation.

Attenuate: Gradually narrowing to a tip or base, this usually narrow and slender.

Bacteria: An infectious single-cell organism that causes disease by breaking down tissue.

Bisexual: Containing both male and female reproductive parts in a flower; a plant that shares both male and female reproductive parts. Having both stamens and pistils.

Bud: The rudimentary state of a stem or branch. Also used for an unexpanded flower.

Bud drop: Abnormal withering of wilting of unopened flower bud, usually from gases of smog and vog, but also from disease or desiccation.

Callous: Having a hard texture, after swollen.

Callus: In cuttings or on injuries, the thick new tissue that develops and covers the injury.

Cambium: A layer of formative cells between the wood and bark in woody plants; the cells increase by division and differentiate to form a new wood and bark.

Chimera: A form of a mutation caused by a DNA change to a portion of the plant so that the visual appearance caused by adjoining tissues of different genetic contitution consists of two (or more) different phenotypes.

Chromosome: A structural body found in the nucleus that is the site of hereditary determinations known as genes.

Clone: Reproduction of an individual from a vegetative part so that it maintains the parental genetics and phenotype. As used in tissue culture, a mericlone is derived from a vegetative meristem.

Corolla: The (normally) most visible part of a flower consisting of petals and tube enclosing reproductive parts. (The outer whorl is the calyx.)

Cortex The portion of the plant stem between the epidermis and the vascular tissue.

Cotyledon: The embryo leaf in a seed, often functioning as a storage organ and the first leaf of a seedling.

Cross-pollination: To remove pollen from a flower of one plant and transfer it to the stigma of a flower of another plant with the goal of creating seeds which will genetically contain superior characteristics of both plants.

Cultivar: A selected hybrid resulting from either asexual or sexual reproduction; horticultural variety or race that has originated and persisted under cultivation, not necessarily referable to a botanical species, and of botanical or horticultural importance, requiring a name.

Damping off: A condition resulting from too much moisture that favors certain fungi that cause rot to the root or basal stem.

Deciduous: Drops leaves and goes dormant annually; falling away, not persistent or evergreen.

Elliptic: Shaped like an ellipse; widest in center and the ends equal. Loosely used. The two drawings show an average example but the leaf can be longer and narrower and still be elliptic.

Embryo: The rudimentary plant within a seed.

Endemic: Confined to a limited geographical area. Synonymous with native.

Epidermis: The outer layer of cells.

Evergreen: Does not go dormant; retains most foliage; continues growing.

Exotic: Not native. Introduced from another area. Compare to indigenous.

Fertile: Capable of producing fruit and seeds; a fertile flower may be pistillate or perfect.

Fertilization: (1) When the egg nucleus of an ovule and the nucleus of a pollen grain unite to form a single-celled zygote. (2) The process of adding nutrients to the soil.

Fetid: With a disagreeable odor.

Filament: Any thread-like body; used especially for that part of the stamen that supports the anther.

Floral envelope: The collective name for the sepals and petals.

Floret: A single flower in an inflorescence; part of the group of flowers.

Floriferous: Bearing flowers, usually many flowers.

Fluted: With grooves or furrows.

Follicle: A dry, dehiscent, one-carpelled fruit with usually more than one seed and opening only along the ventral suture.

Fungus: A growth that includes mushrooms, molds, mildews, and rust which subsist on dead or living organic matter.

Genus (pl. genera): A group of species with certain common, reproducible characteristic.

Gene: One of the determiners of hereditary characteristics, usually located on the chromosomes.

Graft: A means of uniting a separate root system (rootstock, understock) with a shoot system (scion, bud) so that they grow together as one plant. See also grafting.

Hermaphroditic: A flower with both stamens and pistils. Same as perfect and bisexual.

Hybrid: A plant variety different from its parents but with characteristics of both.

Hybridize: To produce or cause to produce a hybrid.

Incurved: Curved toward the axis or attachment.

Indigenous: Native to the area.

Inflorescence: The flowering part of a plant, almost always used for a flower cluster.

Internode: The portion of a stem or other structure between two nodes.

Introduced: A plant brought in intentionally from another area, as for purposes of cultivation. Such a plant may later escape and persist. Compare with exotic.

Inverted: To reverse in position.

Lanceolate: Lance-shaped; several times longer than wide, broadest toward the base, and tapering to apex.

Latex: The milky juice of some plants like milkweed, dandelion, and plumeria.

Leaf: An outgrowth from the stem that is the vehicle for absorbing sunlight for the production of food.

Leaf shapes: Plumerias have many different leaf shapes, which are used for species identification; shapes are described as acuminate, obtuse, obovate, lanceolate, oblanceolate, elliptical, linear, spatulate, and fiddle-shaped.

Meristem: The undifferentiated plant tissue from which new cells are formed, as that at the tip of a stem or root.

Metamorphosis: A profound change in form from one stage to the next in the life history of an organism, as from the caterpillar to the pupa and from the pupa to the adult butterfly.

Microclimate: A small area where climate conditions differ from the general area, sun, shade, air, movement, humidity, etc.

Mutation: A sudden departure from the parent type caused by a change in a gene or chromosome. Also called a sport.

Node: The place on a stem where leaves or branches normally originate; the place on an axis that bears other structures; any swollen or knob-like structure.

NPK ratio: The relative proportions of nitrogen, phosphorous, and potassium in a fertilizer.

Oblanceolate: Inversely lanceolate, attached at the tapered end and broader towards the distal end.

Oblong: Two to four times longer than wide with the sides parallel or nearly so.

Obovate: Inversely ovate, broader above rather than below the middle.

Obtuse: Blunt or rounded at the apex.

Open-pollination: The process of fertilization of the stigma by the introduction of pollen dust by wind or insects, under no control by man, giving unpredictable results.

Ovule: The structure that develops into the seed. It contains one-half of the parental chromosome number in the egg nucleus. The ovary is shown in longitudinal section.

Pedicel: The stalk to a single flower of an inflorescence; also used as a stalk to a grass spikelet. Compare peduncle.

Peduncle: The stalk or solitary flower to an inflorescence. Compare pedicel.

Pendulous: More or less hanging or declined.

Perfect: A flower with both functional stamens and pistils.

Petal: One of the individual parts of the corolla.

Petiole: The stalk to a leaf blade or to a compound leaf.

Phenotype: The visual expression of an organism's genes. See also genotype.

Picotee: Referring to a petal rimmed with a different color, usually pink.

Pistil: The female, ovule-bearing organ of a flower, including the stigma, style, and carpel(s) or ovary containing the ovules.

Plunge: To place a potted plant up to it's rim in soil or certain other material as sand or moss.

Pod: Any dry dehiscent fruit, often used as a synonym for legume.

Pollard: A tree whose top branches have been cut back to the trunk so that it may produce a dense growth of new shoots.

Pollen: The male spores in an anther; fertilizing element. Pollen contains one-half the chromosome number of the plant and its nucleus will unit with the egg nucleus in the ovule to form a zygote.

Pollination: The transfer of pollen from the anther (male part) to the stigma (female part containing the ovule or egg). If fertilization occurs following pollination, then seed is set.

Prostrate: Lying flat on the ground; if a stem, then may or may not root as nodes.

Puberulent: With very short hairs; minutely pubescent.

Reflexed: Abruptly bent or turned downward, or backward.

Root: The usually underground portion of a plant that lacks buds, leaves, or nodes and serves as support, anchoring the plant to the soil. A root draws minerals and water from the surrounding soil, and sometimes stores food.

Rootball: a. The collective mass of roots and soil in the pot of a containerized plant; b. the collective mass of roots and soil attached to the base of a bare-rooted plant.

Rootlet: A small root, often used for the aerial supporting roots put out by some vines.

Rootstock: The root-bearing portion of a plant that is used for grafting. Its crown is replaced by the scion grafted onto it.

Scion: A plant part (detached shoot or twig containing buds from a woody plant) inserted into a rootstock during grafting.

Seed: A ripened plant ovule containing an embryo.

Seedling: A young plant that is grown from a seed.

Seedpod: A dehiscent fruit of a leguminous plant such as the pea, also called seedpod (follicle).

Self-fertilization: The process of pollinating a bisexual floret with its own pollen.

Sepals: The individual greenish structures (collectively, calyx) that are the most basal part of a flower. They may be inconspicuous as in a plumeria or prominent as in cotton.

Shrub: A woody perennial plant smaller than a tree and usually with several basal stems. Compare tree with its drawings.

Species: The basic category of botanical classification which characterizes a group of individual plants of the same ancestry with similar structure and behavior and of stable nature; a group of individuals that has constant, reproducible characteristics that are retained through many generations under natural conditions, usually in isolation. (See also Subspecies, Variety.

Spatulate: Broad and rounded at apex and tapering at base, like a druggist's spatula; flattened spoon-shaped.

Stamen: One of the pollen-bearing organs of a flower; made up of filament and anther.

Sterile: Infertile and unproductive, as a flower without a pistil, a stamen without an anther or a leafy shoot without flowers.

Stigma: That part of the pistil that receives the pollen, usually at or near the apex of the pistil and mostly hairy, papillose, or sticky.

Stock: A rooted plant into which a scion is inserted during grafting.

Style: The usually stalk-like part of a pistil connecting the ovary and stigma.

Subspecies: A variant within a species that is somewhat different from the typical description but not sufficiently different to merit a separate status as a species. Sometimes used when there is only one variant instead of the term "Variety."

Synonym: In taxonomic use, a synonym is the botanical name that is not used because one applied earlier has been accepted as the valid name. Many so-called species of plumeria are actually synonyms.

Taproot: The primary root continuing the axis of the plant downward. Such roots may be thick as in the drawing or comparatively thin.

Tetraploid: having a chromosome number that is four times the basic or haploid number.

Terminal bud: The apical bud.

Terminus: The tip or end of a branch; the point of bloom.

Throat: The opening below the flared petals of the corolla before it narrows into a tube; important for color variants and to lead insects to nectar or pollen.

Tissue culture: The technique of cultivating living tissue in a prepared medium outside the body.

Tuberculate: Bearing small processes, bumps, or tubercles.

Variety: A natural population within a species that shows heritable differences. These differences are not sufficient to give a separate species status.

Virus: An infectious disease that causes growth or flowering defects; to date, viruses are incurable.

Xerophyte: A plant adapted to dry or arid habitats. Compare mesophyte and hydrophyte.

JL seedling

IMAGE INDEX

JL Seedling

INDEX

A

acumulata, 3
acutifolia, 3
air layer, 35
anomalies, 62
anthesis, 70
anthracnose, 53
apocynaceae, 1

B

black sooty mold, 56
borer, 51, 54

C

chimera, 58, 60, 62, 66
clay soil, 45
clone, 119
cold hardy, 45
collecting plumeria, v, 106, 110, 118
color, 1, 3, 4, 5, 6, 7, 9, 10, 15, 17, 26, 32,
    42, 47 , 55, 62, 70, 72, 75, 97, 98, 104,
    107, 118, 123, 125
compact, 4, 16, 22, 97, 73, 74
Conklin, Dean, 116, 117
cross pollinate, 3
cuttings, v, vi, 3, 27, 28, 29, 30, 31, 32, 37,
    38, 39, 43, 47 , 60, 105, 115, 116, 121

D

damping-off, 25
deciduous, 1, 6, 21, 42, 44 , 73, 99, 101,
    108, 122
defoliation, 44
dehydrated, 31, 47
DNA, 58, 119, 121
dormant, 46, 122
dwarf, 7, 28, 30, 54, 73, 74, 75, 96, 97, 99,
    107, 108, 110, 112, 119, 128

E

eBay, 116
emasculate, 71

epidermis, 30, 41, 121, 122
ethephon, 65

F

fertilizer, 25, 31, 37, 40, 41, 43, 45, 47, 56,
    123
form, 4, 16, 28, 47, 58, 121, 122, 123
fragrance, 15, 16, 17, 70, 102, 104, 118
fungi, 25, 30, 53, 54, 56, 62, 122

G

genera, 1, 122
germinate, 25, 69
germinating, 50
gibberellic, 65
girdle, 35
grafting, iii, 32, 33, 34, 60, 122, 124

H

heat lamps, 29, 31
heat mats, 29
hybridize, 71, 72, 122

I

inflorescence, 1, 4, 7, 15, 16, 17, 20, 42, 58,
    59, 109, 118, 122, 123
insects, 51, 53, 55, 56, 69, 70, 71, 123, 125
interiorscapes, 118

J

JL method, 30

K

Koko Head Botanical Garden, 117

L

leaf burn, 44
leaf shapes, 6, 19, 21, 123
light and temperature, 41

Jim Little, photographed by Richard Criley

After receiving an MA degree in art from the University of California at Davis in 1967, Jim Little enjoyed a long and fulfilling career as a freelance photographer and teacher. A treasured highlight from his photography career includes being asked by Bob Goodman, former *National Geographic* photographer, to shoot selected works of Madge Tennet, one of Hawai'i's most reputable artists among Hawaiian women. This project subsequently led to a cover shot in the 1976 issue of *Honolulu* magazine.

Little has taught photography in California and Hawai'i, including Punahou School and the University of Hawai'i. While teaching, he grew plumeria trees as a hobby. In time, Little turned his green thumb and passion for plumerias into a business, exporting plumeria plants to the U.S. mainland and selected international countries. Little, now operating one of the largest plumeria farms in Hawai'i, does research and specializes in the development of new plumeria colors, fragrances, and dwarf forms. With his retirement from Academe, Jim shares his 30 years of experience with and dedication to plumerias in his first book, *Growing Plumeria in Hawai'i*.